PHYSICS AND CHEMISTRY

SECONDARY 2

Enrique Andrés del Río
Francisco Larrondo Almeda
Ángel Rodríguez Cardona
Franciso Martínez Salmerón

W0403264

Pedagogical Advisors:
Juan Carlos Fajardo Gómez
Alicia Sampedro Montañés

Mc Graw Hill Education

MADRID · BUENOS AIRES · CARACAS · GUATEMALA · MEXICO
NEW YORK · PANAMA · SAN JUAN · BOGOTÁ · SÃO PAULO · AUCKLAND
HAMBURG · LONDON · MILAN · MONTREAL · NEW DELHI · PARIS
SAN FRANCISCO · SYDNEY · SINGAPORE · SAINT LOUIS · TOKYO · TORONTO

CONTENTS

HOW TO USE THIS BOOK

Unit introduction

Before you begin, you will explore the key concepts of the unit through a series of **warm-up activities.** Go to the Online Learning Centre (OLC) to download the worksheets for this section.

An interesting **text** will introduce each unit, so that you can see science applied to the real world.

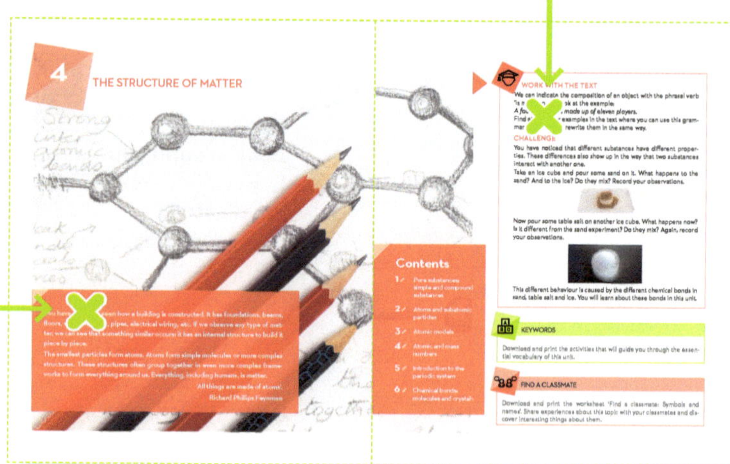

Content development

Learning will be fun with this book. The theory is presented in clear, concise text accompanied by **images** and **infographics.** This visual format will make it easier for you to understand everything.

You will find **different types of activities** to develop different skills, as well as additional information on the side.

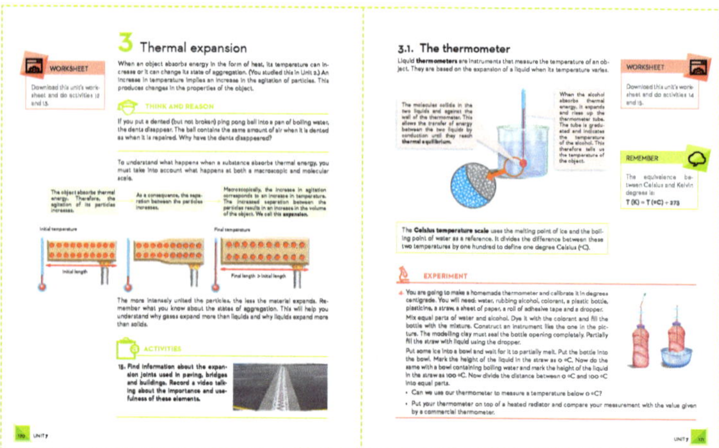

ACTIVITIES

Exercises to consolidate the theoretical concepts you have learned.

SOLVED EXAMPLE

These accompany the theory when the maths is more complex.

THINK AND REASON

Situations to awaken your curiosity and motivate you to learn.

EXPERIMENT

Simple demonstrations you can do with material from home. They only take a few minutes to do and have follow-up exercises.

CLASSROOM LABORATORY

Experiments designed to be performed in the classroom. They will teach you how to resolve the theory through practical tasks.

VIRTUAL SIMULATIONS

Science 2.0 and others. Throughout the book you can access online virtual simulations of the most important physical and chemical processes. You can also access them from the OLC (Online Learning Centre).

WORKSHEET

Complementary activities distributed throughout the different sections of each unit. Download the worksheets from the Online Learning Centre.

End of unit

MIND MAP

Diagram showing the most important concepts in each unit. We invite you to complete them with other concepts you have studied.

LOOK AROUND YOU

Readings, debates and investigations to encourage reflection and critical thinking.

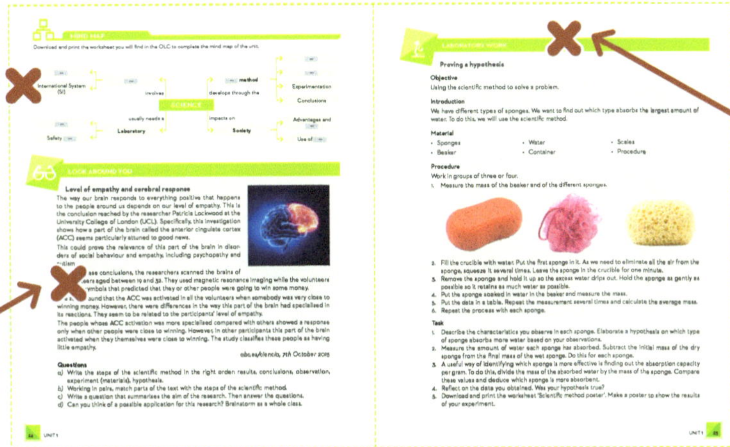

LABORATORY WORK

With related tasks.

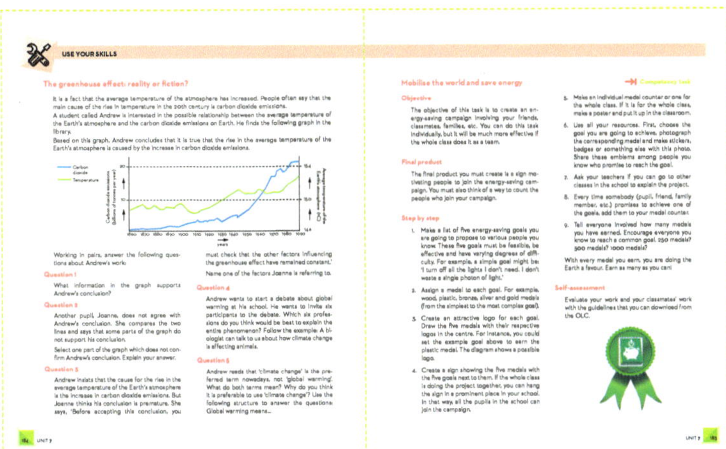

USE YOUR SKILLS

One activity and one task per unit to put your knowledge, skills and abilities to the test.

FINAL ACTIVITIES

Separated into basic, consolidation and advanced activities so you can reach the required level of knowledge.

Research projects

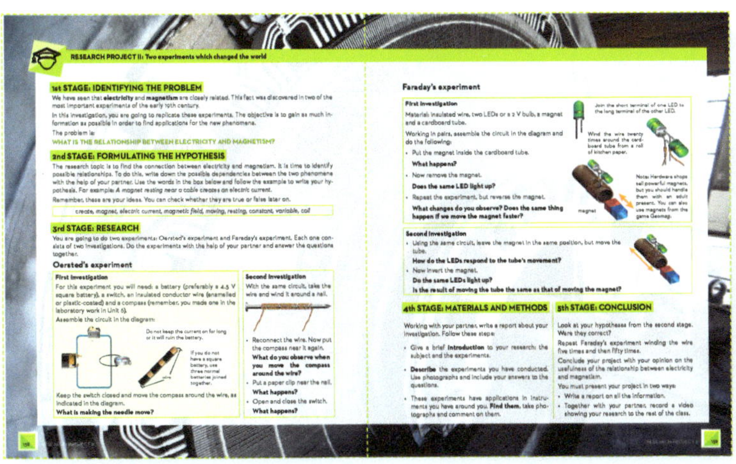

You will work like a real researcher and complete your mission by applying the scientific method.

Remember that you will be asked to go to the OLC at different points throughout the unit to download worksheets that reinforce what you have learnt.

www.mheducation.es

OLC STUDENT

1 SCIENTIFIC ACTIVITY

Judging without knowing. This is very common behaviour in today's society. Sometimes we do not give something that is new or strange a fair chance, often because of fear or ignorance.

Perhaps you have sometimes thought that science is boring. People often think scientists are boring, shut away in a laboratory surrounded by lots of notes. Is that what you think? When you watch television, read the news or listen to the radio, scientists usually appear as people who invent formulas or measure things.

Science is much more than formulas and measurements. In this course you will learn that science is essential in today's society. Scientific advances have changed our way of life.

Have you ever thought about giving science a chance? Immerse yourself in the world of physics and chemistry and perhaps it will surprise you. Learn about them and then you can judge for yourself.

'We have no right to judge what we are incapable of understanding.'

Arthur Rubinstein

WORK WITH THE TEXT

Read the text and write five different sentences starting with the word science. State your opinions and ideas or get the information from the text. The first one has been done for you:

1. Science is not a boring subject at school.

CHALLENGE

You only need to know a little science to do magic. There are very simple tricks that you can do with little preparation. In the past, people used to believe that these tricks were real magic. Here are some simple experiments that you can do:

Turn wine into water!

Take two glasses. Put some water in one of them. Add a few drops of red food colouring to the water to make 'wine' and a small amount of bleach into the 'empty' glass. Pour the 'wine' into the 'empty' glass and describe what happens.

Burning money!

Prepare a 1:1 alcohol-water mixture (the same volume of both) and soak a 5-euro note in it. Use tongs to pick up the note and move it away from the solution.

Light the soaked note with a match. Describe what happens.

a) Prepare a presentation of your 'magic' to an audience.

b) Which branch of science explains your 'magic'? What properties of matter are you using?

c) A lot of 'magic' uses science intelligently. Do you know any other examples?

KEYWORDS

Download and print the activities that will guide you through the essential vocabulary of this unit.

FIND A CLASSMATE

Download and print the worksheet 'Find a classmate: science and job sources'. Share experiences about this topic with your classmates and discover interesting things about them.

Contents

1 Introduction to the scientific method

DID YOU KNOW?

Lots of the words we use in science come from Latin. Learn about the origin of scientific terms:

goo.gl/pJNQwm

Do you like sports? What about playing musical instruments like the violin? Did you know that Albert Einstein loved playing the violin? Or that Edwin Hubble won prizes for athletics and boxing? Or that Ramón y Cajal became a body-builder? Scientists have hobbies just as you do. They are like you.

THINK AND REASON

Niels Bohr was a fan of Westerns, John von Neumann loved social meetings and William Herschel was passionate about music.

a) Do you think hobbies like these are an impediment to becoming a great scientist?

b) Do you think your hobbies could be an impediment to your professional future?

WORKSHEET

Download this unit's worksheet and do activity 1.

In this course we are going to explore the wonderful world of science. But what is science and what is it not? The term 'science' comes from the Latin word *scientia*, which means 'knowledge'. Science can be defined as follows:

Science is a body of knowledge obtained through a set of methods and techniques.

EXPERIMENT **Science is not magic**

1. A long, long time ago, science was called alchemy. Alchemists tried to transform one metal into another. They wanted to change metals such as lead or iron into more expensive metals, like gold. Science as we know it today began in the 19th century thanks to technological advances and a change in perspective.

   When you go home, watch the following video on the history of science: **goo.gl/smVKFJ.** Make a timeline showing the most important scientists and their speciality.

ACTIVITIES

1. This activity consists of a simple search for information. Find the name of five scientists and indicate what they researched. Your description should be brief. Share the results with your classmates. Ask your teacher how long it should be.

2. Identify which qualities all scientists must have. Add a brief description, explaining why you have chosen each quality. For example: perseverance. Scientists work long hours, performing experiments and making observations.

1.1. The scientific method

Science seeks knowledge. To find it, we need a method, a way of searching using a series of ordered steps. This method allows us to identify what is science and what it is not. A scientist uses the scientific method to obtain knowledge.

The **scientific method** is a set of ordered steps that ensure we can explain phenomena with great confidence.

The scientific method has the great advantage of allowing us to retrace our own steps. In summary, it has the following stages:

A. Observation

Scientists are very curious and observant people. Observation is the first stage in the scientific method. At this stage we try to obtain as much information as possible on the issue we want to resolve.

Observation consists of examining a phenomenon with the aim of obtaining as much information as we can without modifying the phenomenon.

ACTIVITIES

3. We want to study the growth of plants from the same species. To do this, we observe factors such as humidity and temperature. What other factors do we have to observe?

4. When we drink cold water, we often get a headache. What information do you need to obtain in the observation of this phenomenon?

SCIENCE 2.0

At the link below there is an amusing video showing the use of the scientific method to solve a problem.

goo.gl/HDzDWh

With a partner, write the observation that starts the research and the different hypotheses tested in the video.

WORKSHEET

Download this unit's worksheet and do activities 2 and 3.

B. The hypothesis

Imagine you are at home waiting for a friend to arrive. She is late and, logically, you start to think about the reasons for her lateness. You look out the window and observe that it is raining and that there are no parking spaces. Is that why? Are there other reasons?

WORKSHEET

Download this unit's worksheet and do activity 4.

The questions you ask yourself are attempts to find an answer to your problem. You do not know if your guesses are true or false. We call these ideas hypotheses.

A **hypothesis** is an idea or explanation for something that is based on known facts but has not yet been proved.

Hypotheses must always refer to real situations, like our example. We must use clear language that says which variables we are connecting.

In this case, when you ask your friend what happened, there are two options: either you were right or you were wrong.

- If you were right, we have a **true hypothesis.**
- If you were wrong, we have a **false hypothesis.**

IMPORTANT

A hypothesis is a possible answer to our problem, but we do not know if it is true or false. Therefore, we must go on to the next stage in the scientific method: **experimentation**.

ACTIVITIES

5. **We express a hypothesis in such a way that we can test if it is true or false. Discuss with your partner what other characteristics a hypothesis must have.**
6. **Which of these two statements serves as a hypothesis? 'Your class is the cleverest one in the whole year.' 'Classes which read more get better marks.'**
7. **Think about the following idea: 'A person's gender influences their intelligence.' Choose your answer (yes/no) and organise a class debate in groups. During the debate, create a diagram and write arguments in favour of both answers. Finally, decide as a group which answer is more likely to be true.**

C. Experimentation

When we think about scientists, we usually imagine people wearing white coats in a laboratory. As you have seen, a scientist is much more than that. Experimentation is the most characteristic and well-known stage of a scientist's work, but do not forget that it is not the only one.

THINK AND REASON

In the example of your friend being late, how can you know if your hypothesis was correct or not?

a) What should a scientist do now?

b) Could you progress to the next step?

Experimentation is a procedure that allows us to test if a hypothesis is true or false.

In this stage we reproduce the phenomenon we want to study. We change the variables and observe what happens. We must be able to modify them in whatever way we want in a safe environment. This allows us to obtain data we can use later.

D. Conclusion

Have you heard of the law of universal gravitation or the theory of relativity? Perhaps you do not know what they say, but you do know that they exist. Why is one a law and the other a theory?

A **scientific law** is a statement that mathematically describes a situation and has been scientifically proven.

A **scientific theory** is an explanation of how and why a phenomenon occurs using different laws and concepts.

If a hypothesis is true, it can become a theory or a law.

Scientists share newly acquired knowledge by publishing their results. This helps us learn about new discoveries. In addition, future researchers can use them as a new starting point for their investigations.

SCIENCE 2.0

The **scientific method** is a process that helps us to find the solution to a problem.

Listen to 'The Scientific Method Song' at the following link:

goo.gl/m2FB4K

List the steps of the scientific method according to the song. Discuss the differences from your textbook with a partner.

WORKSHEET

Download this unit's worksheet and do activities 5 and 6.

ACTIVITIES

8. Brainstorm with the rest of the class to write a list of characteristics that valid experiments must have.

9. Find information about two recent scientific discoveries. If possible, describe what experiments were performed for each one.

10. What experiment would you design to test the following hypothesis? 'Digital watches are more accurate than analogical watches.' Write a step-by-step procedure of your experiment.

IMPORTANT

In science, we consider a property to be a magnitude when we can measure it with a measuring instrument.

WORKSHEET

Download this unit's worksheet and do activities 7 and 8.

2 Magnitudes

Scales or a thermometer are examples of measurement apparatus. We use them to learn the mass of an object or its temperature respectively. Mass and temperature are both examples of magnitudes.

> **Magnitude** is any physical property we can measure.

EXPERIMENT Invent your own unit

2. We are going to take various measurements, but with a new unit. Invent a unit of measurement. Take your own mobile or a pencil. Measure several objects with different sizes, like the desk in your classroom or the blackboard.

   Compare your measurements with those of a classmate who is using a different measuring object. Discuss the problem with this method of measuring things.

To take measurements, we need to establish a numerical value for the property so we can compare results. To do this, we take an internationally recognised quantity as a reference. This is called a **unit**. So, when we say that our dog Rover has a mass of 12 kg, what we are saying is that its mass is 12 times that of the standard unit, which is 1 kg.

> **Measuring** is comparing one quantity with another of the same magnitude that we take as a reference. This reference quantity is called a **unit.**

The dog's mass is twelve times that of the mass set as 1 kg.

ACTIVITIES

11. Indicate which of these properties are magnitudes: happiness, length, beauty, force, niceness, surface, volume, velocity and sadness. Work with your partner to create a table with two columns and think of some other properties to add to this list.

2.1. The International System

On the previous page we asked you to invent a new unit of measurement. We gave you two examples, but there are many more. Therefore, there are lots of different units. But which is the ideal unit?

The **International System of Units (SI)** establishes which unit we use for each magnitude.

The SI aims to provide a common reference so that all countries use the same unit for each magnitude. This means it is easier to work together. The SI establishes the following seven basic units:

Magnitude	Unit in the SI	Abbreviation
Length	metre	m
Mass	kilogram	kg
Time	second	s
Electric current	ampere	A
Temperature	Kelvin	K
Amount of substance	mole	mol
Luminous intensity	candela	cd

IMPORTANT

Fundamental magnitudes are those that we can determine directly, without needing to use a second magnitude. They all appear in the table.

Derived magnitudes are the magnitudes that do not meet this condition.

You will discover these as you study more physics and chemistry.

WORKSHEET

Download this unit's worksheet and do activity 9.

2.2. Multiples and submultiples

Distance is measured in metres in the SI. However, as we can see in these two cases, it is not always helpful to measure in this unit. In science we often find very large or very small numbers. In these cases we use **multiples** and **submultiples** of the SI unit.

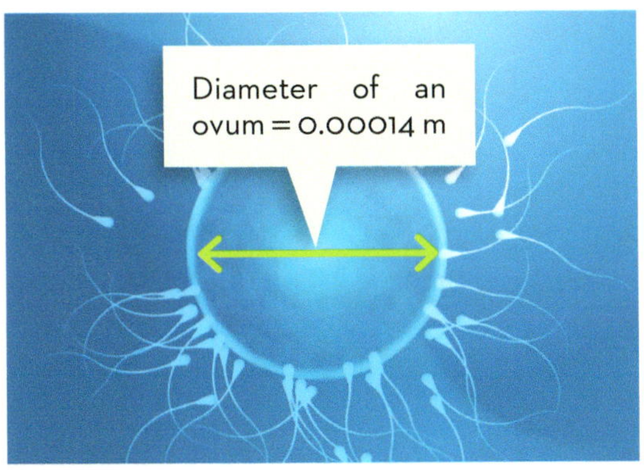

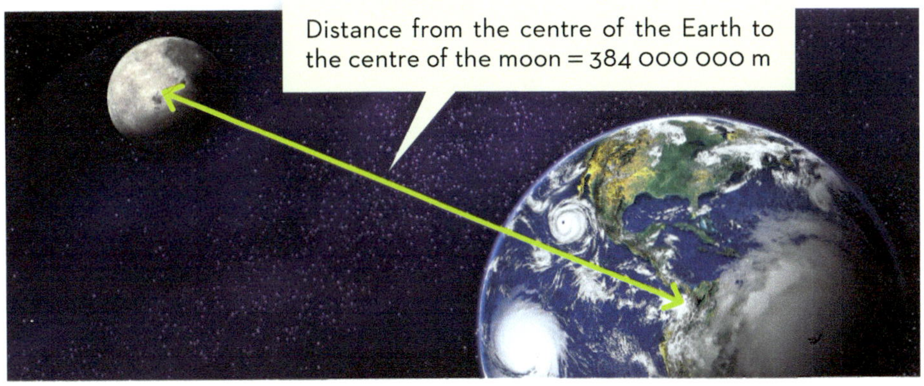

ACTIVITIES

12. Indicate which multiple or submultiple you would use in the following examples: the distance between two cities, the mass of a person, the length of a pencil and the size of an insect.

In science it is common to find amounts that are easier to understand using larger or smaller units. We use multiples and submultiples of base units to express these numbers.

Here are the most common multiples and submultiples:

milli (m)	centi (c)	deci (d)	unit	deca (da)	hecto (h)	kilo (k)
10^{-3}	10^{-2}	10^{-1}		10^{1}	10^{2}	10^{3}

A. Conversion factors

To change units or to use multiples and submultiples, it is a good idea to use conversion factors. This mathematical tool makes these changes easier. Let us look at how to do them.

SOLVED EXAMPLE

1. Convert 340 g to the SI.

Solution

The SI unit for mass is the kg. We have to convert the grams to kilograms.

a) To use conversion factors, we have to multiply by a fraction.

The denominator is the unit we want to convert and the numerator is the unit we want to obtain.

In this case, grams are the denominator and kilograms are the numerator.

Unit I want to convert

$$340\ \text{g} \cdot \frac{\text{kg}}{\text{g}}$$

Unit I want to obtain

b) Now we convert the units. It is a good idea to put a 1 in the larger unit.

Make the conversion: 1 kg = 1000 g

$$340\ \text{g} \cdot \frac{1\ \text{kg}}{1\,000\ \text{g}}$$

Put a 1 in the larger unit

c) Do the mathematical operation. When we divide, we eliminate the grams to give the answer in kilograms, the unit we want.

$$340\ \cancel{\text{g}} \cdot \frac{1\ \text{kg}}{1\,000\ \cancel{\text{g}}} = \frac{340\ \text{kg}}{1\,000} = 0.34\ \text{kg}$$

! IMPORTANT

Other less common multiples and submultiples:

Multiples

Mega (M) → 10^{6}

Giga (G) → 10^{9}

Tera (T) → 10^{12}

Submultiples

micro (μ) → 10^{-6}

nano (n) → 10^{-9}

pico (p) → 10^{-12}

WORKSHEET

Download this unit's worksheet and do activities 10 and 11.

B. Scientific notation

As well as using multiples and submultiples, there is a mathematical tool that allows us to write very large or small values in a simpler way: scientific notation.

Scientific notation is a way of simplifying very large or small written numbers. It uses powers of ten.

SCIENCE 2.0

You can practise conversion factors at the following link:

goo.gl/VSbvdd

SOLVED EXAMPLE

2. Express the following numbers in scientific notation:

a) 156 000 000 *b)* 0.0025

Solution

a) To express a large number using scientific notation, we do the following:

$156\,000\,000 = 1.56 \cdot 10^8$

1. We only put one number before the decimal point.
2. The exponent of ten is the number of places between the first and last numbers.

b) To express a small number using scientific notation, we use a similar technique:

$0.0025 = 2.5 \cdot 10^{-3}$

1. We only put one number before the decimal point.
2. The exponent of ten is the number of places between the first and last numbers. The exponent is a **negative number.**

WORKSHEET

Download this unit's worksheet and do activity 12.

ACTIVITIES

13. Change the following units using conversion factors. Check your answers with a partner.

a) 230 hL to L. *b)* 0.8 h to s. *c)* 5.2 m^2 to cm^2. *d)* 56 mg to g.
e) 6 580 g to kg. *f)* 3.9 cL to daL. *g)* 470 hm^3 to km^3. *h)* 6.25 h to min.

14. Express the following in SI units using conversion factors.

a) 7 dm. *b)* 1.2 h. *c)* 3.4 cm^3. *d)* 4 500 mg.
e) 5 600 mm^2. *f)* 857 000 µm. *g)* 120 min. *h)* 3.2 Gg.

15. Express the following numbers using scientific notation. Check your answers with a partner.

a) 9 000 m/s. *b)* 0.23000 km. *c)* 623 000 min.
d) 0.2350 s. *e)* 600 m. *f)* 0.0006 h.

3 The laboratory

Scientists often work in a **laboratory** (or lab). They perform controlled experiments in the lab to check the hypotheses for a certain problem or circumstance. Let us look at what is in a laboratory.

REMEMBER

Always listen to your teacher. Laboratories are complex places that require experience.

3.1. Instruments

Lots of different instruments are used in laboratories. We are going to describe some so you can familiarise yourself with them.

 Beaker	 Flask	 Test tube	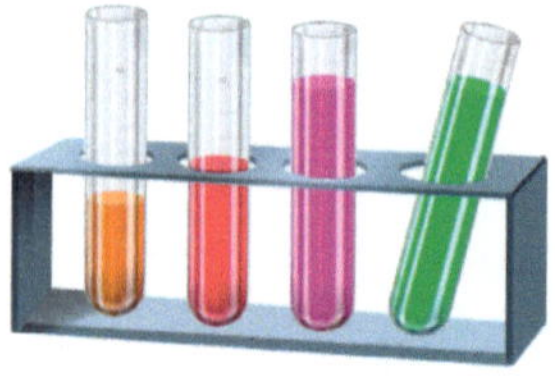 Test tube rack
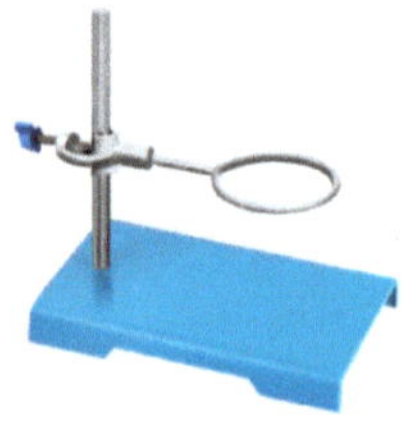 Ring stand	 Microscope	 Graduated cylinder	 Bunsen burner
 Pipette	 Scales	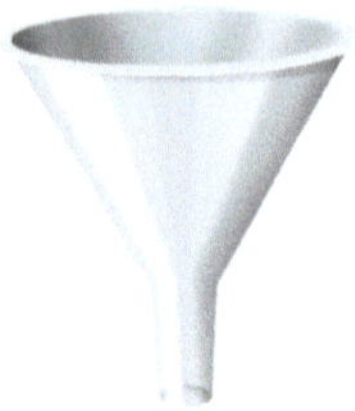 Funnel	 Flat-bottomed round flask
 Erlenmeyer flask	 Safety goggles	 Mortar and pestle	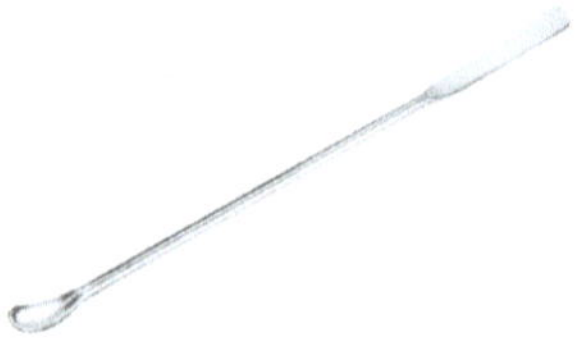 Spatula

CLASSROOM LABORATORY **Learn to use the material**

1. Work in pairs. Your teacher will assign you one of the instruments shown above. Go to your school's lab and count how many there are. Prepare a short report describing what the instrument is used for and the right way to handle it. Present your report to the rest of the class.

3.2. Safety rules

You probably have rules for how to behave at home and at school. This also happens in the laboratory, but the difference is that we work with hazardous products here.

Every laboratory has **safety rules.** Here are the most basic rules that apply to all labs.

EXPERIMENT **Hazard pictograms**

3. All chemical products have labels to show us how hazardous they are. These labels are called hazard pictograms. They appear on everyday products, such as bleach or ammonia.

a) Find information on the meaning of each pictogram and which hazard each one indicates.

b) Find products in your home with these labels. You do not have to find all of them. Indicate which products you have found them on.

c) Watch the video at **goo.gl/aB12Bp**. Identify what the protagonist does with each of the pictograms.

d) Write a short sentence describing the risks taken by the main character as in the example: 'He used a lighter next to an explosive product'.

WORKSHEET

Download this unit's worksheet and do activities 13 and 14.

4 Science in society

SCIENCE 2.0

Read the article you can find at:

goo.gl/Y3T7bp

Write five things that you do every day thanks to scientific advances.

If you ask your parents what their lives were like at your age, they will probably tell you things were very different than they are for you. If you ask your grandparents, the answer will be the same. There are many differences, but there will be one common theme: society.

The evolution of science is an unstoppable fact. We have machines and instruments that make our daily lives easier and that have changed our behaviour. Society has continually changed and adapted with the progress of science.

THINK AND REASON

Consider your normal daily routine, with all your obligations and distractions. Now imagine it without your mobile phone. Would your day be the same? Would you be able to adapt to a lifestyle without it?

Think about how mobiles affect our lives:

WORKSHEET

Download this unit's worksheet and do activity 15.

EXPERIMENT **Scientific advances**

4. Scientific advances are present in our homes. Find appliances or devices which did not exist in your grandparents' time. Ask your parents or grandparents for information. Take a photo of each item and make a data sheet for each one with the following information: year of manufacture, utility, original price when it came on the market and current average price. In groups of four, prepare a poster showing the most important innovations (discuss to decide) and the information you have collected about them.

4.1. Advantages and disadvantages

In Section 4 we talked about mobile phones. You probably noticed that we mentioned both positive and negative aspects. We can deduce that all progress also has its disadvantages. Society needs to progress at the same pace as scientific and technological advances, but this is not always easy.

We are going to reflect on the advantages and disadvantages of scientific progress.

WORKSHEET

Download this unit's worksheet and do activities 16 and 17.

THINK AND REASON

Ask your classmates about the software, websites and social networks that they use most frequently. Make a list with the names of the people you asked and their contributions.

Brainstorm: Can you do the things you do using these tools without the internet? How?

A. The positive contributions of science

Science and technology have brought many benefits. Here are just a few of them:

They facilitate contact between people.

They facilitate the search for information.

They make our work simpler.

They make our lives easier.

They have increased our life expectancy.

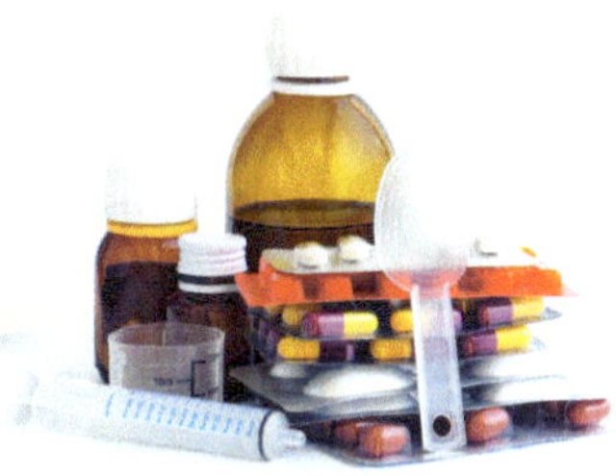

ACTIVITIES

16. **Name five scientific advances in your day-to-day life that have helped you improve your standard of living. Share your ideas in groups of four and decide on the five most important advances. Brainstorm with the rest of the class to make 'The great five' list.**

B. The negative consequences of how we use science

We cannot forget that some scientific advances also have their disadvantages. Here are just a few of them:

New products: increase in the amount of waste produced.

Communication: bullying and harassment.

Transport: drones and the resulting loss of privacy.

ACTIVITIES

17. Read the following article:

The scientific advance that threatens the future of humanity according to Hawking

'Success in creating artificial intelligence would be the biggest event in human history. Unfortunately, it might also be the last unless we learn how to avoid the risks.' This is the latest warning from Stephen Hawking. It could be them or us.

Intelligent robots 'have the capacity to improve themselves and be smarter than us,' says the scientist in an interview with the American channel HBO. An extreme which could lead to a confrontation between machines and humans. The result is obvious to Hawking: 'We would lose.'

Hawking recognises that the potential benefits of this technological development could be 'huge', but he warns that, in the short term at least, they are headed more towards self-destruction. 'Artificial intelligence could eradicate war, disease and poverty in the world. However, we can see how priority is given to developing weapons.'

Another argument Hawking presented in an opinion article for the newspaper *The Independent* is that 'the short-term impact of artificial intelligence depends on who controls it, the long-term impact depends on whether it can be controlled at all.' These advances are happening very fast and society does not have time to react.

The Independent, 1st May 2014

a) Make a diagram showing pros and cons of artificial intelligence according to the article.

b) Do you agree with Mr Hawking? Give reasons for your point of view.

c) In groups organise a debate about the advantages and disadvantages of artificial intelligence. Search for information on other scientists' opinions to learn about different points of view. Use them in your debate.

4.2. Science and ICT

Do you know what a digital blackboard and a tablet have in common? What about an internet blog and an online magazine? They all have something in common: they are ICT.

ICT stands for **Information and Communication Technology**. They are tools which allow a new way of communicating through the use of new technologies.

Education today is very different than it was in the past thanks to these tools. They provide us with more ways to receive information.

IMPORTANT !

ICT tools are the result of scientific progress. They allow us to communicate with one another and to receive information very quickly.

CLASSROOM LABORATORY ICT in your classroom

2. Follow these steps:

   a) Observe your classroom.
   b) Find everything you consider to be ICT in the classroom.
   c) Make a note of them and write a brief summary of their uses.
   d) Approximately how many times a week do you use each of these tools?
   e) What do you like most about ICT?

Online magazines have created a new form of information. They are much more accessible and convenient than printed magazines.

Tablets provide us with a convenient way of accessing information. These devices are more manageable than laptops.

WORKSHEET

Download this unit's worksheet and do activities 18 and 19.

ACTIVITIES

18. Find information about three online scientific magazines. Write down their website addresses and briefly comment on each website.

5 Your first research project

ICT allows us to access large amounts of information in a simple way. It is time for you to use ICT to do your first research project.

WORKSHEET

Download this unit's worksheet and do activity 20.

A **research project** is a procedure based on the scientific method. It allows us to gather information on a phenomenon and formulate a hypothesis to explain it.

We are going to give you some advice to guide you during your first project. Remember, we always follow the steps of the scientific method.

ACTIVITIES

19. Do your first research project. A group of students wants to discover if the amount of salt that can be dissolved in the same amount of water depends on temperature. Formulate a hypothesis and follow the steps outlined above. Present your work in a document that must have the following sections: problem, hypothesis, testing procedure (materials), results, conclusions and references.

5.1. Veracity of the information

Fortunately, there are many sources where you can find information. But are they all reliable?

THINK AND REASON

A friend tells you about a news story. You did not know this news until they told you.

a) Do you believe your friend whatever the story is?

b) Imagine several people told your friend this story. Would you be more certain that the news story was true?

c) Now imagine that you have read the same story on the internet, but it is not consistent with what your friend told you. Which of the two sources of information would you believe?

When searching for information, never forget to think critically. Search for information in science magazines such as *Science* **(www.sciencemag.org)**, *Investigación y ciencia* **(www.investigacionyciencia.es)** or *Nature* **(www.nature.com)**.

IMPORTANT

We must never forget to use **critical thinking.** We receive lots of information from the media, but it is not always accurate.

WORKSHEET

Download this unit's worksheet and do activities 21 and 22.

ACTIVITIES

20. Read the following article:

86% of people support banning a highly dangerous chemical substance

An American high school student decided to demonstrate how susceptible humans are, especially when we are given information about the dangers of chemical products. He asked 50 people to sign a petition to ban a certain widely used chemical substance. He cited the following risks to our health:

1. It can cause excessive sweating and vomiting.
2. It is a main component of acid rain.
3. It can cause serious burns in its gaseous state.
4. Accidental ingestion of this product has caused many deaths.
5. It significantly contributes to the erosion of the planet.
6. It reduces the efficiency of brakes in vehicles.

43 out of the 50 people signed in favour of the government banning this substance.

Answer the following questions:

a) Debate as a group. How many people in the class would sign the petition?

b) Working in pairs, do research to find out chemicals that present the risks listed in the text.

c) Can water be the chemical substance the text regards as highly dangerous? Discuss with your partner.

d) The chemical substance that has these characteristics is water. Now that you know this, would you sign the document?

e) What conclusion can you draw from this article?

MIND MAP

Download and print the worksheet you will find in the OLC to complete the mind map of the unit.

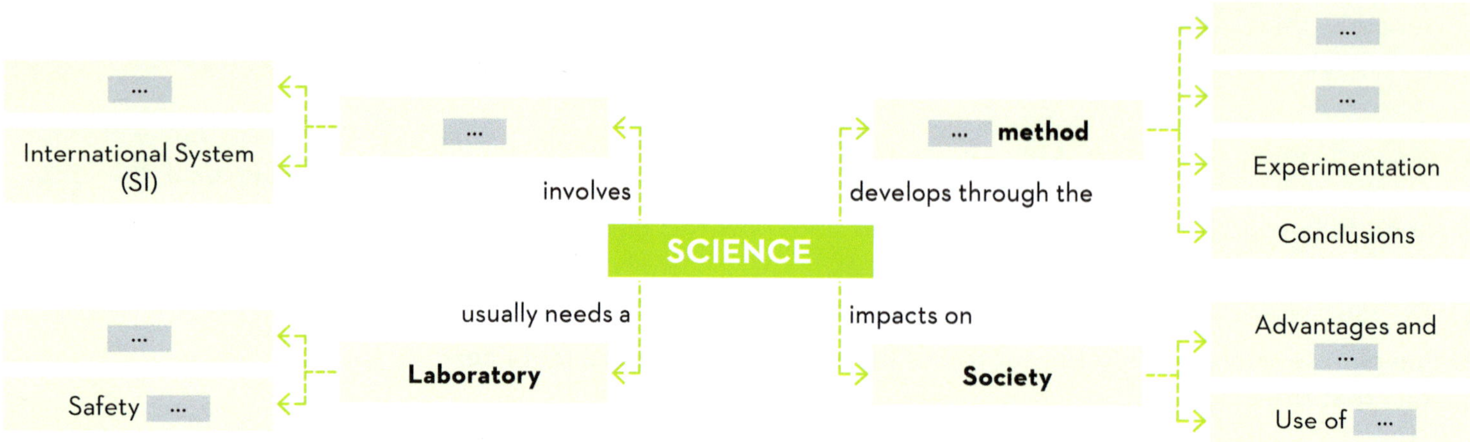

LOOK AROUND YOU

Level of empathy and cerebral response

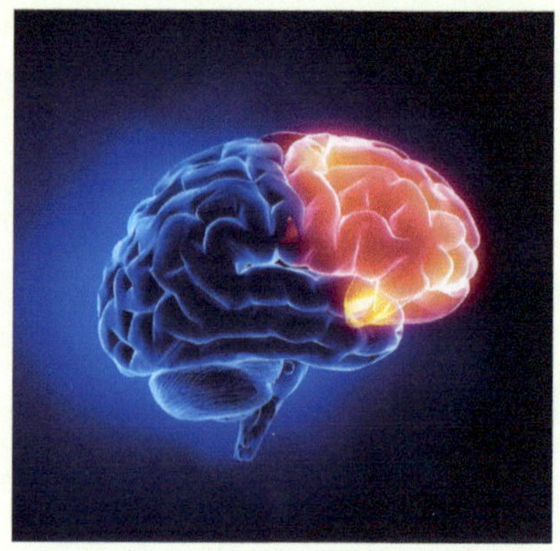

The way our brain responds to everything positive that happens to the people around us depends on our level of empathy. This is the conclusion reached by the researcher Patricia Lockwood at the University College of London (UCL). Specifically, this investigation shows how a part of the brain called the anterior cingulate cortex (ACC) seems particularly attuned to good news.

This could prove the relevance of this part of the brain in disorders of social behaviour and empathy, including psychopathy and autism.

To reach these conclusions, the researchers scanned the brains of 30 volunteers aged between 19 and 32. They used magnetic resonance imaging while the volunteers looked at symbols that predicted that they or other people were going to win some money.

The study found that the ACC was activated in all the volunteers when somebody was very close to winning money. However, there were differences in the way this part of the brain had specialised in its reactions. They seem to be related to the participants' level of empathy.

The people whose ACC activation was more specialised compared with others showed a response only when other people were close to winning. However, in other participants this part of the brain activated when they themselves were close to winning. The study classifies these people as having little empathy.

abc.es/ciencia, 7th October 2015

Questions

a) Write the steps of the scientific method in the right order: results, conclusions, observation, experiment (materials), hypothesis.

b) Working in pairs, match parts of the text with the steps of the scientific method.

c) Write a question that summarises the aim of the research. Then answer the questions.

d) Can you think of a possible application for this research? Brainstorm as a whole class.

LABORATORY WORK

Proving a hypothesis

Objective

Using the scientific method to solve a problem.

Introduction

We have different types of sponges. We want to find out which type absorbs the largest amount of water. To do this, we will use the scientific method.

Material

- Sponges.
- Beaker.
- Water.
- Container.
- Scales.
- Procedure.

Procedure

Work in groups of three or four.

1. Measure the mass of the beaker and of the different sponges.

2. Fill the crucible with water. Put the first sponge in it. As we need to eliminate all the air from the sponge, squeeze it several times. Leave the sponge in the crucible for one minute.
3. Remove the sponge and hold it up so the excess water drips out. Hold the sponge as gently as possible so it retains as much water as possible.
4. Put the sponge soaked in water in the beaker and measure the mass.
5. Put the data in a table. Repeat the measurement several times and calculate the average mass.
6. Repeat the process with each sponge.

Task

1. Describe the characteristics you observe in each sponge. Elaborate a hypothesis on which type of sponge absorbs more water based on your observations.
2. Measure the amount of water each sponge has absorbed. Subtract the initial mass of the dry sponge from the final mass of the wet sponge. Do this for each sponge.
3. A useful way of identifying which sponge is more effective is finding out the absorption capacity per gram. To do this, divide the mass of the absorbed water by the mass of the sponge. Compare these values and deduce which sponge is more absorbent.
4. Reflect on the data you obtained. Was your hypothesis true?
5. Download and print the worksheet 'Scientific method poster'. Make a poster to show the results of your experiment.

FINAL ACTIVITIES

Basic activities

1. Read the following article:

A boy was lost on a desert island, abandoned by its former inhabitants. As it was cold, he looked for material to make a fire. Then he carried the things he found to his camp. He discovered that some of them burned but others did not. The boy wrote two lists.

Things that burn: tree branches, broomsticks, pencils, chair legs, flag poles.

Things that don't burn: rocks, marbles, paperweights, old plates.

When the boy had burned nearly all the tree branches and broomsticks, he tried to find a regularity that would serve as a rule. Looking at his classification, he noticed an apparent relationship between the objects. He thought of a possible 'generalisation'. Perhaps cylindrical objects burn.

The next day he deliberately left his list at his camp. Guided by his rule this time, he returned with his arms full. He had three pieces of pipe, two plastic bottles and the axle of an old car. He did not bring a large cardboard box full of newspapers.

During the long, cold night, the boy reached the following conclusions:

Perhaps the cylindrical rule is unrelated to combustibility. However, the broomsticks, the tree branches and the bowling pins are cylindrical and they burn.

He observed the list again and, after some reflection, proposed the following rule. Wooden objects burn.

Decisively, he went to look for the wooden door and threw it on the fire. His satisfaction was tremendous. Then he thought:

Dry wood burns better than wet wood. Once wood burns, the residue does not burn again. It is getting harder to find wooden objects each day. This is worrying. I must look for other objects that burn. Tomorrow I will try walnut shells and some roots I discovered today.

Chemistry: An Experimental Science
CHEMProject

Answer the following questions:

a) What was the boy's first hypothesis?

b) What experiment did he do to discard his first hypothesis?

c) Do you think he used the scientific method? Explain your answer.

d) What was his final hypothesis?

2. Indicate which of the following properties are magnitudes: flavour, bitterness, temperature, length.

3. Discuss with a classmate to decide if the following sentences are true or false:

a) Alchemy is a branch of science.

b) A hypothesis is an unproven conjecture.

c) The scientific method is a cyclical method.

d) There are magnitudes which cannot be measured.

4. Change the following units using conversion factors:

a) 20 mm to m.
b) 85.2 g to hg.
c) 0.062 m^2 to cm^2.
d) 52.5 hm^3 to m^3.
e) 526 cm to dam.
f) 5 h to s.
g) 520 min to h.
h) 8000 mm^2 to cm^2.

Consolidation activities

5. Express the following measurements in SI units:

a) The mass of a 150 g object.

b) The volume of a 333 cm^3 can.

c) The size of a 370 cm^2 newspaper.

6. When we cut an apple, we observe that the exposed part turns brown after a short time. Form a hypothesis about what happens. (Remember, it does not have to be true.) What experiment could you do to prove or disprove your hypothesis?

7. Express the following in SI units using conversion factors:

a) 56 g. *b)* 120 min. *c)* 0.26 hm^2.

8. Express the following numbers in scientific notation:

a) 28 000. *b)* 0.000569. *c)* 0.02365.

9. Indicate the hazard pictogram for each of the following statements:

a) Poisonous substance when ingested, inhaled or comes into contact with skin, even in small doses.

b) Highly reactive substance which can destroy live tissue.

c) A substance which has a harmful effect on the environment.

Work with a partner. Invent a hazard pictogram and explain its meaning.

10. Indicate which safety rules are not being followed in the image. Indicate what consequences there may be from not following these rules.

Advanced activities

11. The following units do not belong to the International System: inch, ounce, tonne, pound, hectare, gallon and fathom. Find information about each one and classify them into units of length, mass, surface or volume.

12. Express the following in SI units using conversion factors:

a) 2 000 000 pm.

b) 5.1 Gm.

c) 560 000 ng.

13. Julia has measured the temperature of her bedroom at 30-minute intervals since 9 o'clock in the morning. She has obtained the following results: 10 °C, 12 °C, 15 °C, 16 °C, 20 °C, 23 °C, 25 °C. Compile the data in a time-temperature table and then draw a graph, putting time on the X-axis.

14. Working as a group, choose an important scientific advance from the 20th century and prepare a short presentation on it. Give reasons for your choice.

15. In the laboratory you want to prepare a solution of ethanol in water, using 100 mL of water and 20 mL of ethanol. You have at your disposal: both liquids, laboratory equipment and the help of your teacher.

a) Indicate what lab equipment you need to use.

b) Indicate what steps you will follow to obtain the solution.

16. At the following link **goo.gl/hEajhY** there is a video to help you reflect on scientific advances:

a) Discuss the content of the video, focusing on the issues of using technology at home.

b) Describe a situation that shows the effects of the misuse of technology on social relationships.

USE YOUR SKILLS

Colony collapse disorder

An alarming phenomenon called **colony collapse disorder** (CCD) is threatening beehives around the world. It is also known as beehive depopulation syndrome.

CCD occurs when bees abandon the beehive. Separated from the beehive, the bees die and this problem has caused the death of thousands of millions of bees. Experts believe there are several factors responsible for the phenomenon.

One possible cause for CCD is the insecticide **imidacloprid**, which can make the bees lose their sense of direction when they are outside the beehive.

The experts have done tests to see whether exposure to the insecticide causes the syndrome. They added the insecticide to the bees' food in some beehives for three weeks. They exposed several beehives to a different concentration of the insecticide while other beehives were not exposed to any insecticide.

None of the beehives depopulated immediately after exposure to the insecticide. However, after 14 weeks some of the beehives had been abandoned.

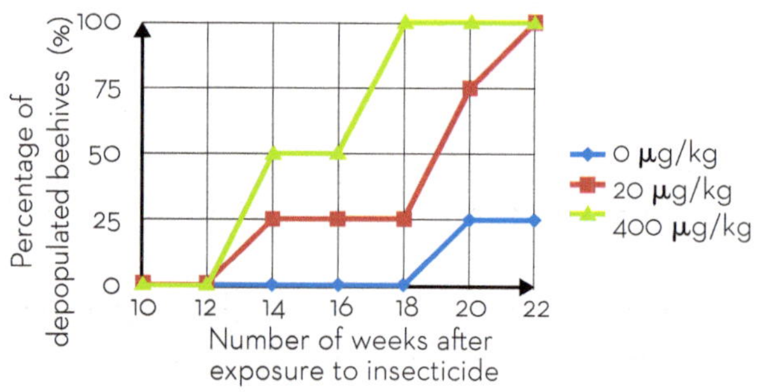

Question 1

This syndrome does not just affect bees. Ornithologists have also observed the phenomenon in birds. Sunflowers are a food source for both birds and bees: bees feed on the nectar of the flower while birds eat the seeds.

Is there a relationship between the disappearance of bees and the population of birds? Discuss in groups.

Question 2

a) Where did the researchers test the effect of the insecticide?

b) What did they modify to perform the tests?

Question 3

Analyse the graph and decide which statements are true. Change the false statements to true sentences.

a) The beehives exposed to a higher concentration of insecticide tend to depopulate sooner.

b) The beehives exposed to the insecticide tend to depopulate after an exposure period of ten weeks.

c) The beehives exposed to the insecticide do not survive more than fourteen weeks.

Question 4

Identify the steps of the scientific method used in this research and draw your conclusions. Discuss these conclusions in small groups.

Truth or lies?

→| Competency task

Objective

Our objective is to reflect on scientific rigour in the media and how easy it is to manipulate information. The media inundates us with scientific news stories about medicines, statistics or new advances. We tend to believe these stories whatever they say, because they often deal with subjects we are not familiar with. We are going to demonstrate this with a news item we have invented and people's response to it. This will show us if we are susceptible to these news stories or not.

Final product

There are various ways of doing this task. However, we think the best option is a classroom presentation. (You can use any software you like.) Analyse the scientific rigour of the news item and the answers people give to your questions.

The aim is to observe how easy it is to manipulate scientific information and communicate it to the general public because they do not have enough scientific knowledge.

Step by step

Here is one possible sequence of steps:

1. Invent a news story. This is the most difficult part and we recommend you ask your teacher for help.

   You can also find fake stories on the internet. The news item should be short and current. For example: the discovery of a medicine that cures all diseases, the discovery of a conspiracy among industries to dispose of radioactive products illegally, etc. Try not to make it too fantastical, as it must be believable and short. A short news story will reach more people.

2. Develop a series of questions about your story. Remember that people have to answer your questions. They should be short and direct questions about the news item.

3. This is the fun part. You will have to go out and ask people to read the story and answer the questions. Do this task in groups. One person could ask the questions while another writes down the answers. This also helps give the story greater credibility. When they have finished, tell them it was a school project and that the news story was not true. Always thank them for their collaboration. Try to interview at least ten people.

4. Indicate how many people believed the story and how many did not. Calculating the percentage is sufficient.

5. Present your task in class. To do this, create a presentation which includes your news story, the questionnaire and people's answers. In addition, express your personal opinion about the results. The aim is to generate a debate, so ask your classmates about the results obtained.

Self-assessment

Evaluate your work and your classmates' work using the guidelines that you can download from the OLC.

2 THE PROPERTIES OF MATTER

The Greek philosophers asked themselves the question 'What is matter?' thousands of years ago. It all began with the philosopher Democritus around 400 BC and reaching the current definition of matter has not been easy. It took thousands of years to establish one.

At the end of this unit you will understand what matter is and you will be able to identify what is matter and what is not matter. This understanding will serve as a foundation for you to gradually increase your scientific knowledge. This knowledge will be useful in your daily life.

Car windows fogging, water evaporating in a pan and ice cubes are three examples of the phenomena we will study in this unit. As you go through the lessons, you will find the first scientific theory of this course.

Effort and patience are two characteristics shared by all scientists. You may find this unit quite difficult, but do not worry, your efforts will be rewarded. Remember, science is worth the effort.

'The doer alone learneth.'

Friedrich Nietzsche

WORK WITH THE TEXT

In the text on the left, there are many words that are derived from others. For example, *musical* derives from *music*. Find as many derived words as you can in the text and write the word they come from.

CHALLENGE

We are going to create a small rainbow of colours in a glass. You will need: syrup, honey, sunflower oil, olive oil, water and alcohol.
We are going to mix everything in a champagne glass using a spoon.

1. Pour some syrup in the glass, using the spoon to help you.
2. Add a little honey.
3. Add a little water.
4. Add sunflower oil and then olive oil.
5. Finally, pour in some 96% alcohol.

Answer the following questions:

a) Describe the arrangement of the liquids in the glass. Use the words 'top' and 'bottom' as in this example: Alcohol is on top of the other liquids.

b) Try adding the liquids in a different order and describe what happens.

c) Explain the arrangement of liquids using comparative sentences as in the example:
Olive oil floats on water because it has a smaller density.
Syrup sinks below water becaue it has a greater density.

d) Select some items from around the house (safety pin, macaroni, jelly bear...) and drop them carefully in the champagne glass. Describe what happens.

KEYWORDS

Download and print the activities that will guide you through the essential vocabulary of this unit.

FIND A CLASSMATE

Download and print the worksheet 'Find a classmate: guess the word'. Share experiences about this topic with your classmates and discover things about them.

Contents

1 Matter and its properties

DID YOU KNOW?

In ancient Greece, the philosophers believed that all matter was made of four elements: earth, water, fire and air. This supposition did not explain the properties of matter.

Physics and chemistry study matter. They study its internal composition as well as its appearance and external properties. Therefore, it is important to understand what matter is. In this way, we can identify its characteristics and learn to differentiate it from other phenomena.

THINK AND REASON

- Can matter be invisible?
- If I cannot touch it, is it matter?

Are you sitting in a chair? Do you have a desk in front of you? If we tell you they are matter, you probably will not be surprised. However, it is more difficult to determine if other things such as air, fire or light are matter or not.

For something to be classified as matter, it needs to fulfil two conditions:

Matter is everything that has mass and occupies space.

EXPERIMENT Scientific knowledge

1. Our objective is to find out what people know about the concept of matter. Working in pairs, to do this, follow these steps:
   a) Go out onto the street and ask people if the following substances are matter or not: earth, water, fire and air. Write down the results.
   b) Draw a table showing the results and indicate which answers are the most common. We recommend you use percentages so you can compare the results.
   c) Explain what matter is to your friends and family. Then do the same survey with them.
   d) Is there a difference between the two sets of answers?
   e) Aristotle thought that all matter was made of these four elements. Are they matter or not? Now it is your turn to answer this question.
   f) Write a brief report on the results and conclusions.

ACTIVITIES

1. **State which of the following substances are matter: happiness, a building, sand, fire, oxygen, water, sadness.**

1.1. The properties of matter

Your desk has a certain size, colour and smell. In other words, it has a series of properties. All matter has properties which characterise it. We can classify all these properties into two groups: general and specific.

A. General properties

Mass and volume are typical examples of the general properties of matter. What do these two properties have in common?

General properties are the ones which all matter has. The value of these properties does not inform us about the type of matter.

Look at these two identical glasses with the same amount of liquid (200 mL, for example) in them. One of the liquids is water and the other is ethanol.

Volume does not tell us which liquid is in each glass. Therefore, **volume** is an example of a general property.

SCIENCE 2.0

Find more information on the properties of matter. The video embedded in this website shows examples of the properties of matter and how to classify them:

goo.gl/J4Eznr

WORKSHEET

Download this unit's worksheet and do activities 1 and 2.

B. Specific properties

Think about your desk again. Imagine you paint it with a mixture of colours. The colour of your desk would be unique and only your desk would have this exact mixture of colours in the whole world. Unlike general properties, this property would allow us to distinguish your desk from any other similar desks.

Specific properties are the ones with a unique value for each type of matter. This property allows us to identify what something is.

Two examples of specific properties are the density and boiling point of a substance. For instance, the boiling point of water is 100 °C. There is no other substance in the world that has this value, only water.

ACTIVITIES

2. **You have two glasses containing different liquids. You have been told that one of them is water, but you do not know what the other liquid is. Design an experiment that allows you to differentiate between them. Say which property you are referring to.**

3. **Give three examples of general properties and three examples of specific properties. Use different examples from the ones mentioned in this section. Share your list with a partner. Did you think of the same properties? Try to make a larger list together.**

1.2. Mass

Mass is one of the essential properties that a substance must have to be considered matter.

> SCIENCE 2.0
>
> You can find exercises to practise converting measurements of mass and volume at the following link:
>
> **goo.gl/1fnjNZ**

THINK AND REASON

We have a piece of wood, a piece of metal and a bottle containing water.

a) Do you know which of the three has more matter?

b) Can we measure this quantity?

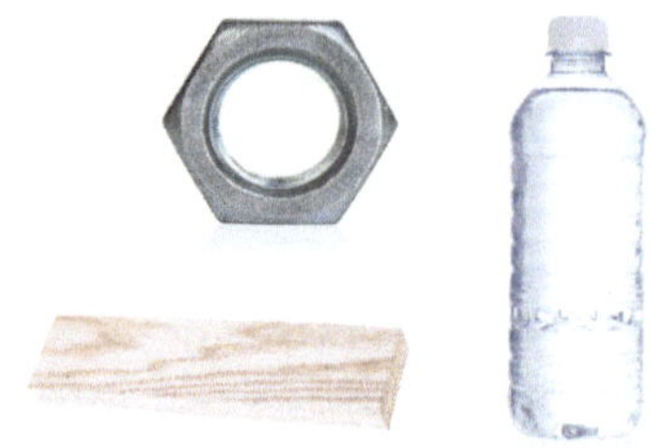

We can define mass in the following way:

Mass is a measurement of the quantity of matter an object has. The SI unit for mass is kilograms (kg).

> WORKSHEET
>
> Download this unit's worksheet and do activity 3.

The mass of a substance is independent of temperature, pressure or location. For example, an astronaut with a mass of 80 kg has the same mass when he is on Earth as when he is on the moon. The quantity of matter is the same.

To measure mass we use instruments such as **scales**.

1.3. Volume

Volume is the other essential property a substance must have to be considered matter.

Volume is a measurement of how much space an object occupies. The SI unit for volume is cubic metres (m^3).

Therefore, an object with a greater volume means that it occupies more space. Volume is independent of shape. For instance, if we have 1 L of water and distribute it between several glasses, we still have the same amount. Therefore, it occupies the same amount of space.

SOLVED EXAMPLE

1. We normally measure volume in litres (L), even though the SI unit is m^3. Do the following conversion: 500 cm^3 to L.

Solution

We will use the equivalency: $1\ dm^3 = 1\ L$

$$500\ \cancel{cm^3} \cdot \frac{1\ \cancel{dm^3}}{1000\ \cancel{cm^3}} \cdot \frac{1\ L}{1\ \cancel{dm^3}} = 0.5\ L$$

We calculate the volume of an object differently depending on its shape. If it has a regular shape, such as a sphere or a cylinder, we can calculate its volume mathematically with the appropiate formula.

$V = \pi \cdot r^2 \cdot h$

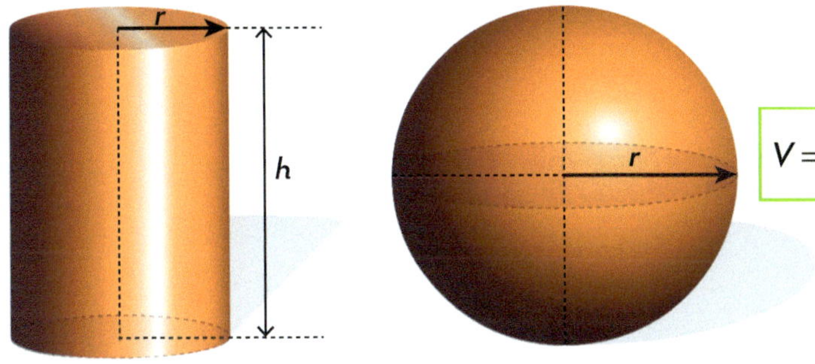

$V = \frac{4}{3} \cdot \pi \cdot r^3$

However, there are not many regular shapes in nature. Most objects have an irregular shape, like a stone or our own bodies. How can you calculate your own volume?

THINK AND REASON

Imagine you have a bath filled to the brim with water and you climb into it.

a) What happens to the water?

b) Once your whole body is in the bath and you sink to the bottom, does the same thing happen?

In these cases there is no mathematical formula to help us. We use a fundamental principle of science: two objects cannot occupy the same space. When we submerge an object in liquid, the volume of the liquid increases. The volume of the object we immerse in the liquid is exactly equal to the increase in volume.

SCIENCE 2.0

Learn how to do different types of volume measurements with this animation:

goo.gl/P5c7Bn

WORKSHEET

Download this unit's worksheet and do activities 4 and 5.

CLASSROOM LABORATORY **The volume of a stone**

1. Working in pairs, you are going to determine the volume of an irregular object. You will need: a graduated cylinder, water and a marble stone.
   a) Fill the graduated cylinder to the halfway mark with water. Write down the water level.
   b) Put the marble stone into the water and write down the new volume.
   c) The difference between the two volumes is the volume of the marble stone.
   d) Create a table to show the results of your experiment. Repeat with an object that does not sink.

ACTIVITIES

4. Convert the following units:

a) 500 g to kg.

b) 250 cm^3 to L.

c) 1200 dg to Hg.

d) 1950 m^3 to mL.

5. Calculate the volume of a sphere with a 10 cm radius. Give your result in SI units.

2 Density

You have learned about the concepts of mass and volume and their respective units. For all matter, we can relate these two concepts through a new property we call density.

Density is the ratio between the mass and the volume of an object.

$$d = \frac{m}{v}$$

Since we measure mass in kg in the SI and volume in m^3, we measure density in kg/m^3. This indicates how much matter in kg there is in 1 m^3 of a substance. It is a specific property and therefore each type of matter has a unique density value.

DID YOU KNOW?

The Dead Sea is situated between Jordan and Israel. It is the lake with the highest concentration of salt in the world. It has so much salt that you can float in it in almost any position. This is because of the very high density of the water.

THINK AND REASON

Is it easier to float in a swimming pool or in the sea? Base your answer only on the density of water in both cases. Density of seawater = 1027 kg/m^3. Density of water = 1000 kg/m^3.

Why does a stone sink in water but oil floats on it? To answer this question, we must look at the density of the three substances. A stone sinks because it is denser than water; oil floats because it is less dense than water.

EXPERIMENT **Distinguish between normal and zero cola**

2. To do this experiment you will need: a can of normal cola, a can of zero cola and a receptacle containing water.
   a) Slowly put the can of normal cola into the receptacle. Does it sink or float?
   b) Now put the can of zero cola into the same receptacle. Does it sink or float?
   c) How can you explain what just happened?

WORKSHEET

Download this unit's worksheet and do activities 6 and 7.

ACTIVITIES

6. Calculate the density of an object with a mass of 2.3 kg that occupies a volume of 1.2 m^3.

7. Slowly pour sunflower oil and olive oil into the same glass. Find the density of both liquids and explain what happens. Write a short procedure for this experiment, indicating the steps and results.

SOLVED EXAMPLE

2. The density of mercury is 13 600 kg/m³. Calculate the volume that 200 grams of this metal would occupy.

Solution

To solve these problems, we must be careful with the units.

We have to convert the unit of mass to kilograms because that is the unit we use for density. That way, we can obtain the volume in m³:

$$m = 200\,\cancel{g}\cdot\frac{1\text{ kg}}{1\,000\,\cancel{g}} = 0.2\text{ kg}$$

$$d = \frac{m}{V} \longrightarrow d\cdot V = m \longrightarrow V = \frac{m}{d} = \frac{0.2\,\cancel{kg}}{13\,600\,\frac{\cancel{kg}}{m^3}} = 1.47\cdot 10^{-5}\text{ m}^3$$

SCIENCE 2.0

We can find a practical application of the use of density in the story of Hiero and the gold crown. Discover the story at:

goo.gl/c1CCG7

If you only want the part about density, go to:

goo.gl/y7LrxX

EXPERIMENT **Ice and water**

3. Do ice and water occupy the same amount of space?

We are going to answer this question with an experiment. You will need: a plastic receptacle (to prevent breakage) and water.

1. Fill the container with water to a certain point and use a marker pen to mark the level of the water.
2. Put the receptacle in the freezer. Wait for two hours and take it out of the freezer. Is the ice at the same level as the pen mark?
3. Based on the result, do ice and water have the same density?

In order to calculate density we need to measure the mass and the volume of a substance.

- To measure **mass**, we use scales.
- To measure **volume**, we use a graduated cylinder or a similar instrument. This way, we can measure the increase in volume when we put an object in water.

ACTIVITIES

8. **Find information about substances with the highest and lowest densities in the world. Make a presentation comparing their densities and their uses in everyday life.**

9. **We have a graduated cylinder containing 20 cm³ of water. We put an object with a mass of 280 g into the water. It sinks and the level of the water rises to 25 cm³. Calculate its density. Compare your results with a partner.**

10. **Calculate the density of an object with a mass of 500 g which displaces the water in a graduated cylinder by a volume of 220 cm³. Compare your answers with a partner.**

3 States of aggregation

SCIENCE 2.0

Read about the states of matter at:

goo.gl/T8yXwx

Take the test to check your understanding. Record your score.

Water exists in liquid form in the seas and oceans, in gaseous form in the atmosphere or in solid form in icebergs.

As we can observe, matter can present itself in different forms in nature.

> A **state of aggregation** is the form in which matter appears in nature.

Matter can present itself in three different forms, each one with its own characteristic properties.

State		
Solid	Liquid	Gas
	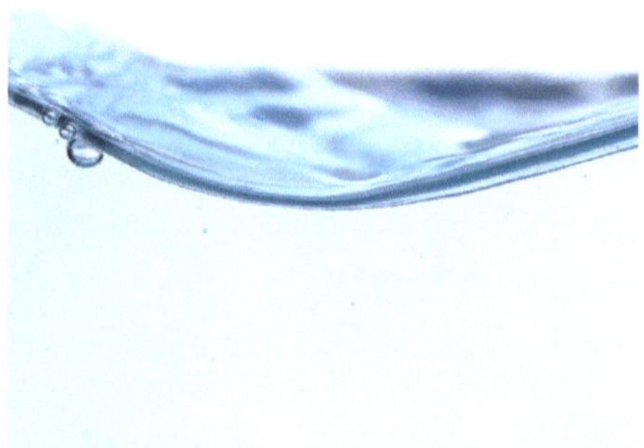	
Shape		
Constant	Variable. It adapts to the receptacle.	Variable. It adapts to the receptacle.
Volume		
Constant	Constant	Variable. It adapts to the receptacle.

ACTIVITIES

11. We have 2 L of a gas in a closed receptacle. We put it in a closed room and take off the lid.

a) What happens to the gas?

b) Will its volume be different now? Discuss your answer with a partner.

12. Plasma is considered to be the fourth state of matter. Find information on its use and in what conditions it is stable. Prepare a short presentation on the topic.

WORKSHEET

Download this unit's worksheet and do activities 8 and 9.

3.1. Changes of state

A state of aggregation is stable within a certain range of pressure and temperature. However, if we change these conditions, we can convert one state of matter into another. For example, melting a block of iron at a high temperature.

A **change of state** is a variation in the form in which matter appears, but its composition remains the same.

Therefore, we have the same type of matter but it is organised in a different form. Each change of state has its own name, as shown in the diagram below:

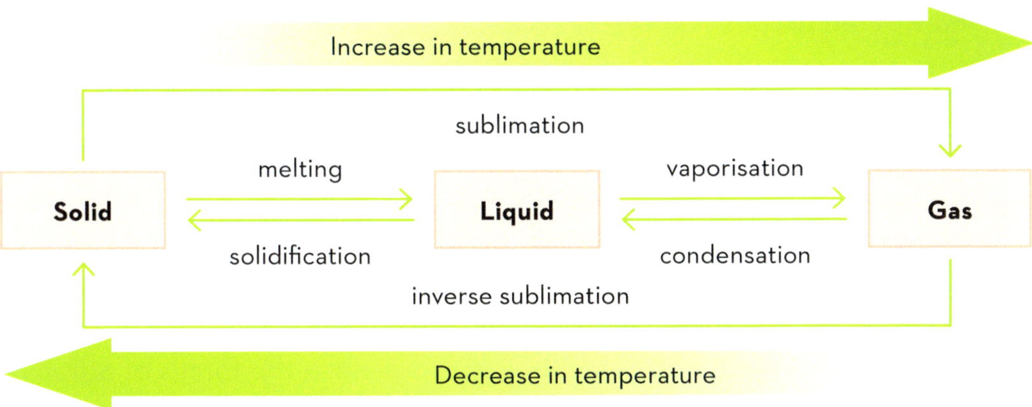

As you can see from the table, the change of state from liquid to gas is called vaporisation. This process can occur in two ways.

- **Evaporation**: This affects the surface of the liquid and happens at any temperature. One example is when wet clothes dry.
- **Boiling**: This affects the entire liquid and happens at a specific temperature. One example is when we heat water to 100 °C. This specific temperature is the boiling point.

EXPERIMENT Sublimation

4. Sublimation is an unusual change of state, but we can use the element iodine to experiment with it.
   1. Put a certain amount of iodine in a beaker.
   2. Gently heat it in a well-ventilated area.
   3. Iodine vapours begin to form. When this happens, cover the beaker with a previously chilled watch glass. What can you observe on the part of the watch glass that is in contact with the vapours?

ACTIVITIES

13. Indicate what type of transformation occurs in the following situations:

a) Gel air freshener. *b)* Mist forms on the bathroom mirror.

SCIENCE 2.0

Learn as you play! This is an application where you can manipulate the changes of state and their characteristics:

goo.gl/UWgtkP

WORKSHEET

Download this unit's worksheet and do activity 10.

SCIENCE 2.0

In Unit 4 we will study the particles mentioned in the kinetic theory of matter and prove they really exist. You can learn more about this theory on this website. It has animations and information about real-life applications of the kinetic theory:

goo.gl/AhpJea

Click on the Union Jack icon for the English version.

WORKSHEET

Download this unit's worksheet and do activities 11 and 12.

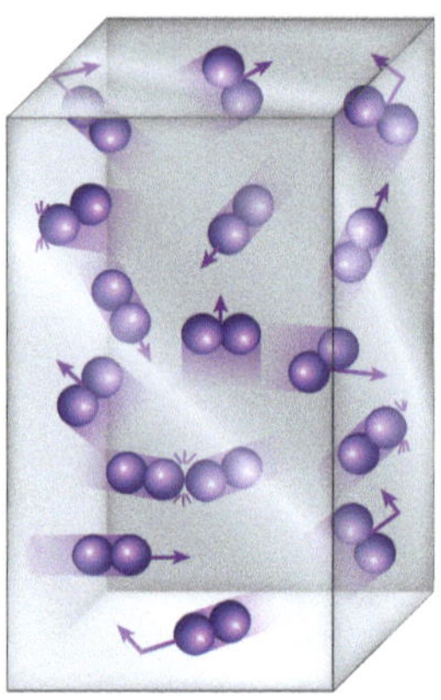

4 The kinetic theory of matter

What is the internal structure of matter? How can we explain the properties of each state? Scientists asked these questions and found the answers to them a long time ago.

This is the first theory in our adventure in physics and chemistry: the kinetic theory of matter.

The **kinetic theory of matter** consists of three postulates which explain the behaviour of the different states of aggregation. A postulate is an undemonstrated proposition that is accepted.

FIRST POSTULATE

Matter is made of very small, indivisible particles. There is nothing between them, only empty space.

SECOND POSTULATE

There are forces of attraction between the particles. These forces keep the particles together. The intensity of the attraction makes the difference between solids, liquids and gases. In a gaseous state, the forces of attraction are negligible. In a solid state, they are so strong that they prevent the particles from moving apart.

THIRD POSTULATE

Particles are in continual, chaotic motion. They move randomly and without stopping.

THINK AND REASON

The diagram represents what a gas would look like if we could see its particles. What would happen to these particles if they did not find a barrier to stop their movement?

The particles in a gas move freely at high velocities. This explains why gases **do not have a fixed shape or volume.** They completely fill the container they are in.

The particles move until they find something that stops them. The force of attraction between these particles is very weak. This explains their capacity to expand.

ACTIVITIES

14. If we spray perfume in a room, we can smell it shortly afterwards. Explain this fact using the kinetic theory. Discuss your answer with a partner.

4.1. Justification of the states of aggregation

The kinetic theory was developed to describe the behaviour of gases. However, it also serves to explain the properties of other states.

A. The properties of solids

We can justify the characteristics of the solid state thanks to the kinetic theory. To do this, we are going to perform the following experiment:

EXPERIMENT **Construct a solid state**

5. We are going to build a model of a solid in accordance with the kinetic theory. Take a transparent jar and fill it with marbles. Try to fill it as much as possible to simulate the solid state effectively.

   The jar represents a solid and the marbles represent the particles in the solid.

   According to kinetic theory, there is space between the particles. Can you observe this in your solid?

According to the kinetic theory, solids have particles with strong attraction forces holding the particles close together, like the marbles in the jar. This means that solids have a **fixed shape and a definite volume**. There is space between the particles. The particles vibrate continually, but stay in their position of equilibrium. The same happens when you shake the jar of marbles quickly.

B. The properties of liquids

As with solids, the liquid state can be explained with the kinetic theory.

Look at this picture of a play area with balls. The balls simulate the particles of a liquid and the net is the receptacle.

If you break the barrier enclosing the balls, will the final volume be the same as the initial one?

According to the kinetic theory, the force of attraction between particles is weaker in liquids than in solids. This explains their capacity to flow (change shape easily). However, they do not have the capacity to expand indefinitely. Therefore, they have a **fixed volume** but **no definite shape**.

DID YOU KNOW?

Plasma is considered to be the fourth state of matter and it is achieved by heating a gas to a very high temperature.

Science is constantly searching for new states of aggregation and ways to make use of their properties.

ACTIVITIES

15. What relationship is there between the melting point of a solid and the strength of the bonds between its particles?

4.2. Temperature and pressure

A. Temperature

We are made up of millions and millions of particles in continual motion. Can we measure this movement?

DID YOU KNOW?

Thermometers are the instruments we use to measure temperature. The sale of mercury thermometers has been banned since 10 April 2014 because it is a hazardous substance.

EXPERIMENT **Recovering its shape**

6. We are going to repair a dented ping pong ball using temperature.

   Dent a ping pong ball and put it in a receptacle containing hot water. Move the ball with a fork or a spoon so the entire ball heats up.

   a) What happens?

   b) How can you explain this using the kinetic theory?

WORKSHEET

Download this unit's worksheet and do activities 13 and 14.

To find a way of measuring the motion of particles, we need to define a new magnitude: temperature.

Temperature is a measurement of the internal energy of a substance. It is related to particle motion.

When the temperature increases, the energy of the substance is greater. Physically, this translates into a **faster movement of the particles.**

There are various scales to measure temperature. The most common scales are the **Celsius scale** and the **Kelvin scale.**

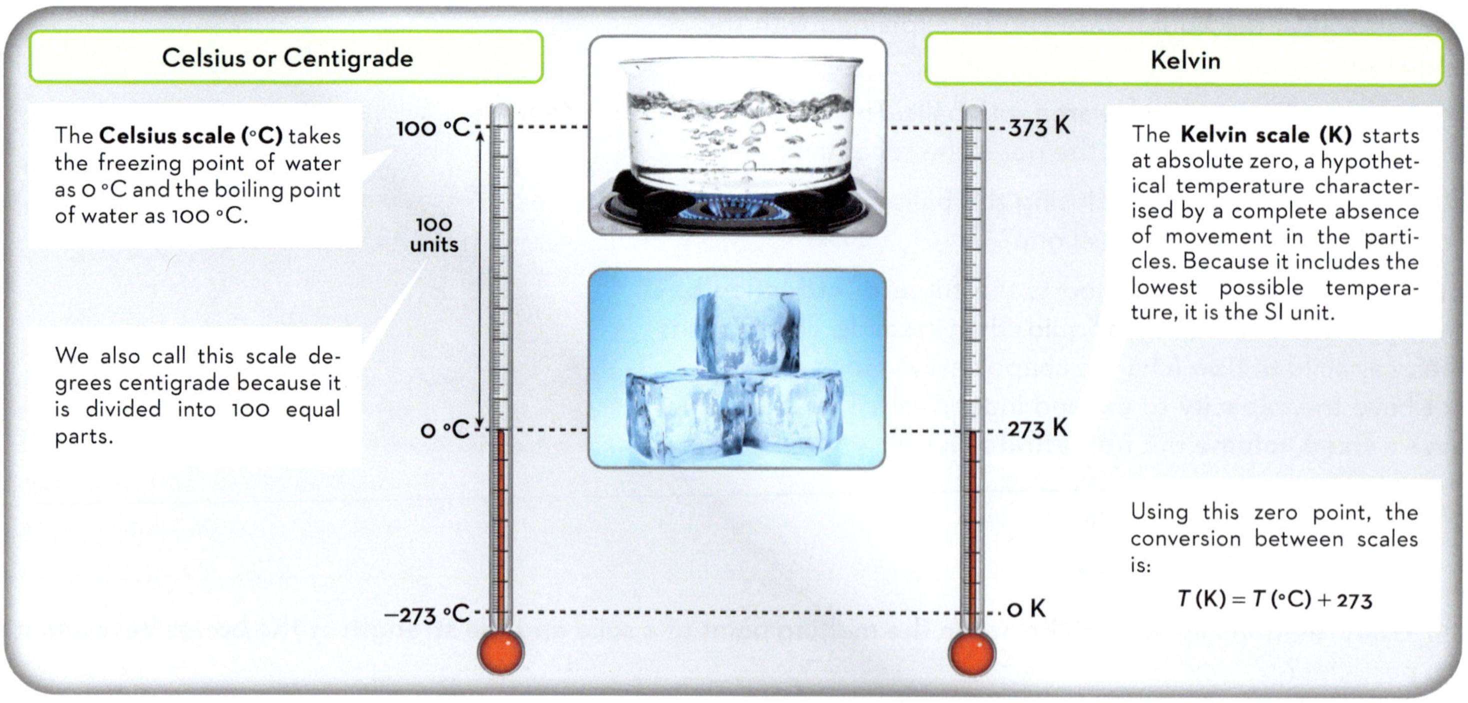

B. Pressure

The particles of a gas continually collide against one another and against the walls of a receptacle. Can we measure this in some way?

THINK AND REASON

Imagine some basketballs represent the particles. You and a couple of friends are in the middle of a room, bouncing the basketballs off the walls again and again.

a) Do these collisions have any effect on the walls?

b) If you increase the number of collisions, would it have a greater effect?

Do you know why balloons sometimes burst? The answer is in the kinetic theory. When particles repeatedly collide against the walls of a receptacle, they can sometimes overcome the barrier that contains them. In the case of the balloon, the particle motion can break the walls. We can measure these collisions with another magnitude: pressure.

Pressure is a measurement of the number of particle collisions against the walls of a receptacle.

We measure pressure in **pascals, Pa** in the International System. The more pressure there is, the more collisions there are and vice versa.

We can also measure pressure in **atmospheres, atm**. 1 atm is the atmospheric pressure at sea level. To convert between the two units, we need to know the equivalency:

$$1 \text{ atm} = 101\,325 \text{ Pa}$$

DID YOU KNOW?

Learn more about atmospheric pressure. At this website you will find self-assessment exercises and animations:

goo.gl/d1ds8h

WORKSHEET

Download this unit's worksheet and do activity 15.

CLASSROOM LABORATORY **Implosion**

2. We are going to cause an implosion. You will need: a fizzy drink can, water, a heat source, a receptacle and tongs.
   1. Partially fill the can with water and heat it (using a candle, for example).
   2. Hold the can with tongs and quickly turn it upside down over a receptacle containing very cold water. Observe what happens.

   What is an implosion and why does it occur?

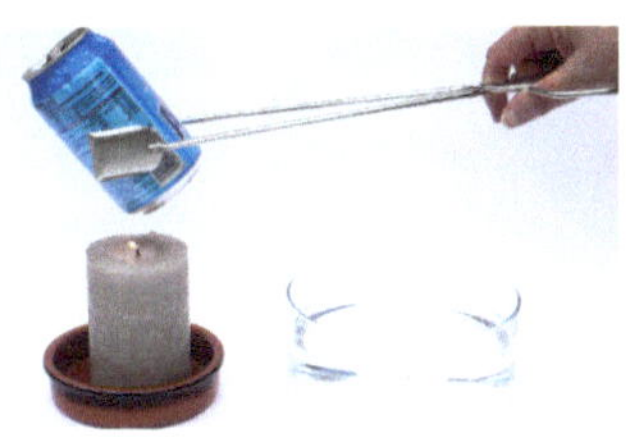

ACTIVITIES

16. Do the following conversions:

a) 20 ºC to K. *b)* 450 K to ºC. *c)* −2 ºC to K. *d)* 0 K to ºC.

17. Do the following conversions:

a) 2.3 atm to Pa. *b)* 200 000 Pa to atm. *c)* 0.23 atm to Pa.

4.3. Changes of state

The kinetic theory does not only explain the states of aggregation in matter. It also explains how matter changes from one state to another.

A. From solid to liquid

If we have a solid at a certain temperature and we increase the temperature, the particles do not change position, but they do vibrate faster.

At a specific temperature, the energy supplied is enough to separate the particles from each other. We say the substance has changed to a liquid state. We call this temperature the **melting point**.

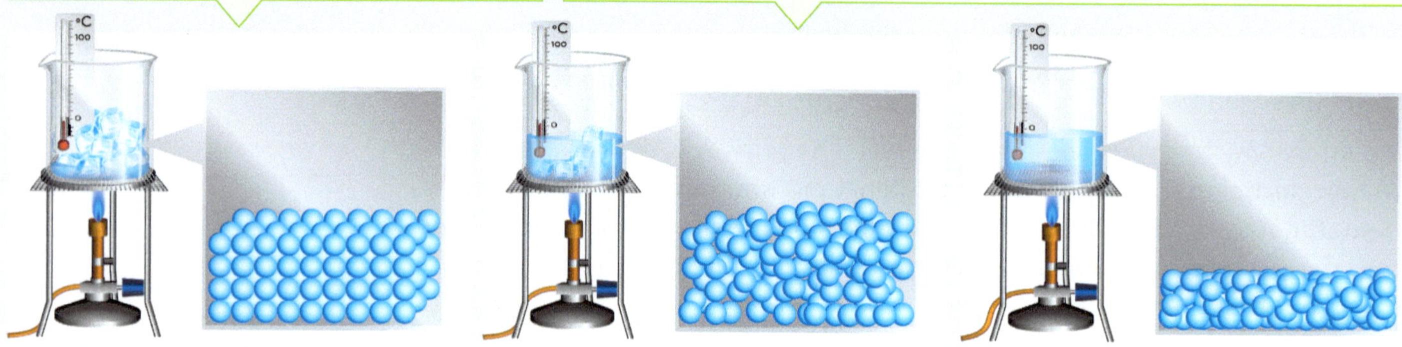

B. From liquid to gas

In a liquid state, particles have a greater capacity for movement. If we heat the liquid, we increase the particle motion.

At a certain temperature, the particles have gained so much energy that they can separate from each other, changing into a gaseous state. We call this temperature the **boiling point**.

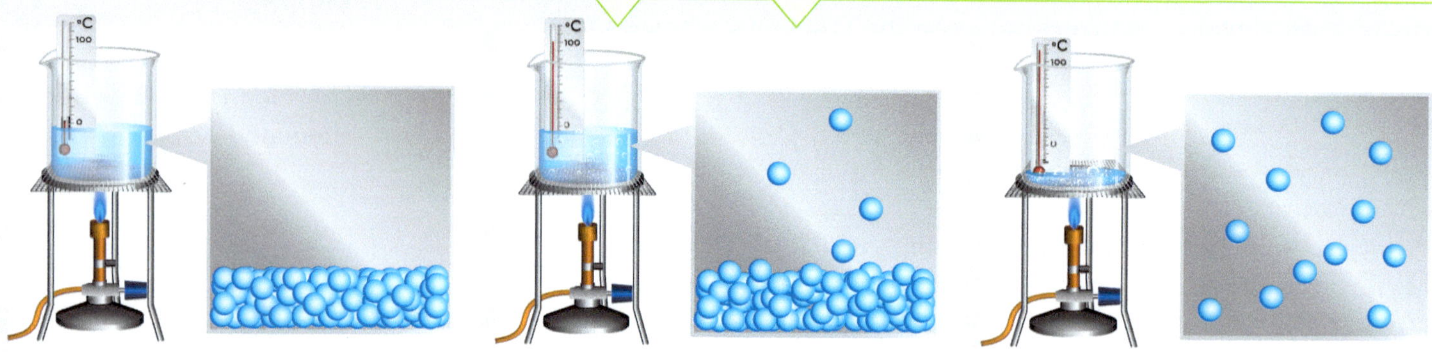

As we saw earlier, evaporation and boiling are two distinct concepts.

- **Evaporation** is a phenomenon in which the surface of the liquid converts into a gaseous state. This can happen at any temperature.
- **Boiling** occurs when the entire liquid converts into a gaseous state. This happens at a specific temperature.

EXPERIMENT **How temperature changes**

7. Put some water in a saucepan. Put a thermometer in the water and heat it. You have to be patient and observe the thermometer. Observe how the temperature changes over time.
What happens to the thermometer when the water starts to boil? What about when all the water is boiling?
Graph the temperature vs. the heating time. Do you notice any change in the shape of the graph?

C. Graph showing changes of state

It is useful to draw a graph showing the temperature values a substance acquires when we heat it until a change of state occurs.

A **heating curve** is a graph showing the increasing temperature of a substance and the time it takes for a change of state to occur.

A heating curve is a very useful tool. It tells us the range of temperature in which the state of aggregation of a substance is stable.

Heating curves have two characteristic areas:

- The **inclined lines** show us where the temperature varies. Therefore, the substance is in a state of aggregation and its temperature is rising.
- The **horizontal lines** are periods of constant temperature. Therefore, a change of state is occurring: there is a transition between two states. Two states of aggregation coexist in these periods.

SCIENCE 2.0

This animation shows the heating curve for water and how to do it step by step:

goo.gl/a1tHgx

SOLVED EXAMPLE

3. The graph below shows the heating curve (time and temperature) for ethanol. Answer the following questions:

a) What is the range of temperature in which ethanol remains in a liquid state?

b) What is the boiling point of ethanol?

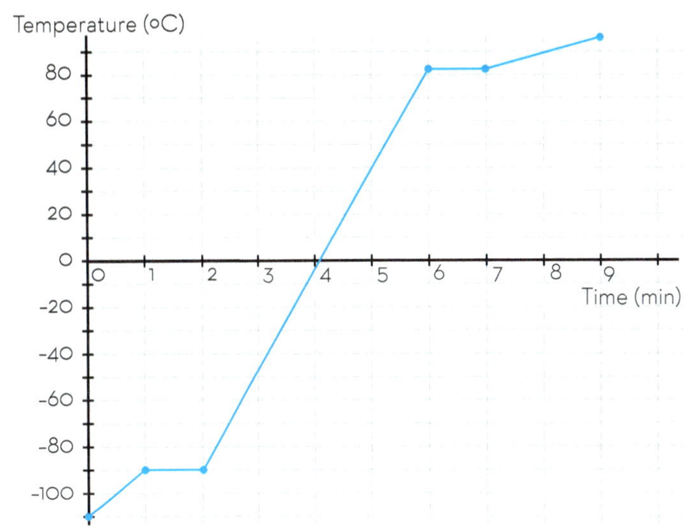

Solution

a) As we can see from the graph, ethanol remains in a liquid state between −90 °C and 80 °C, approximately. At higher temperatures, it is in a gaseous state; at lower temperatures, it is in a solid state.

b) The graph shows that ethanol begins to change from a liquid to a gas after 6 minutes. Its boiling point is approximately 80 °C.

ACTIVITIES

18. Draw the graph from Solved Example 3 in your notebook. Answer the following questions:

a) Mark the solid, liquid and gaseous states on the graph.

b) Indicate the melting point for ethanol.

c) Compare your graph with a partner and check your answers with section b.

WORKSHEET

Download this unit's worksheet and do activities 16 and 17.

5 Gas laws

We are going to take a closer look at gases, one of the three states of matter. As we have learned, this state of aggregation is characterised by three variables: pressure, volume and temperature.

SCIENCE 2.0

In this website you will find an easy animation that illustrates the gas laws:

goo.gl/tkfaFK

THINK AND REASON

Scientists originally developed the kinetic molecular theory to explain the behaviour of gases.

Why do think they used this state and not the liquid or solid states?

Our objective is to find out what relationship exists between these three variables. We are going to keep one variable fixed and observe how the other two variables behave.

The **gas laws** are a set of experimental laws explaining the relationship between the pressure, volume and temperature of a gas with a constant mass.

Each law is named after the scientists who performed experiments to define them.

WORKSHEET

Download this unit's worksheet and do activities 18 and 19.

5.1. Boyle-Mariotte's law

The Irishman Robert Boyle and the Frenchman Edme Mariotte discovered this law. Those physicists conducted their experiments at a constant temperature to observe how volume and pressure relate to each other.

The **Boyle-Mariotte Law** says that pressure and volume are inversely proportional for a certain amount of gas at a constant temperature.

$$P_1 \cdot V_1 = P_2 \cdot V_2$$

To obtain this result they performed experiments at a constant temperature. They modified the pressure to observe the variation in volume. They recorded the data in a table and drew the corresponding graph. Then they searched for a relationship between the two magnitudes and found one.

Volume (L)	Pressure (atm)
25.2	2.5
18	3.5
12.5	5
9	7

ACTIVITIES

19. Imagine you are one of the two physicists from Section 5.1. Use the data in the table to draw a *P-V* graph. Put pressure on the horizontal axis.

a) Looking at the results, were the scientists right?

b) Calculate the volume at a pressure of 10 atm.

c) Compare your graph with a partner and check your answers with *b)*.

EXPERIMENT Expand a marshmallow

8. You will need: insulation tape, a large 60 mL syringe and a marshmallow.
   1. Put the marshmallow inside the syringe.
   2. Seal the narrow top of the syringe with insulation tape. If you do not have any, cover it tightly with your hand.
   3. Move the plunger slowly up and down. What happens to the marshmallow with each of the two movements?

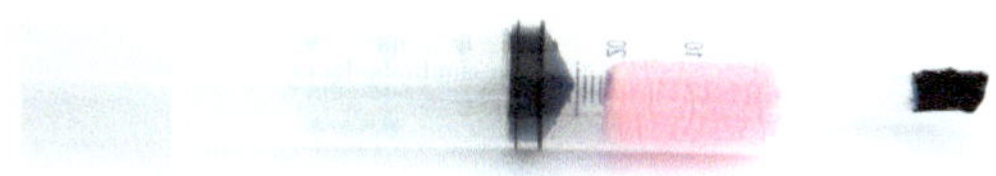

To explain this law, we must turn to the kinetic theory once again.

We have a gas enclosed in a receptacle. If we push the plunger down, **we reduce the volume**. The gases have less space to move around and therefore collide against the walls of the receptacle more often. This causes an **increase in pressure**.

SCIENCE 2.0

You can change the pressure and volume in this application to test how they affect each other according to Boyle-Mariotte's Law:

goo.gl/y1Roab

SOLVED EXAMPLE

4. We have a gas enclosed in a cylinder at a pressure of 1.2 atm, occupying a volume of 2.5 L. We increase the pressure to 2 atm and the temperature is constant during the process. Calculate the final volume.

Solution

We are working at a constant temperature, so we apply Boyle-Mariotte's Law. We enter the data in the mathematical equation:

$P_1 = 1.2$ atm $\quad P_2 = 2$ atm $\quad V_1 = 2.5$ L $\quad V_2 = ?$

$$P_1 \cdot V_1 = P_2 \cdot V_2 \rightarrow 1.2 \text{ atm} \cdot 2.5 \text{ L} = 2 \text{ atm} \cdot V_2 \rightarrow V_2 = \frac{1.2 \cdot 2.5}{2} \rightarrow V_2 = 1.5 \text{ L}$$

WORKSHEET

Download this unit's worksheet and do activity 20.

ACTIVITIES

20. We have a gas at a pressure of 1.6 atm. We increase the pressure to 3 atm, obtaining a volume of 200 mL. We have maintained a constant temperature. Calculate the initial volume of the gas.

21. At a constant temperature, how does the pressure change if the volume doubles? Look for the appropiate verb to indicate the change in pressure.

5.2. Charles's law

Jacques Charles discovered this law in 1787. The French scientist studied the relationship between the temperature and the volume of a gas at a constant pressure.

> **DID YOU KNOW?**
>
> Using his law, Charles predicted what we call **absolute zero**. He said that this is the lowest temperature that gases can reach and its value is **−273 °C**.

THINK AND REASON

We have some marbles in a bag. The bag can become as large as we want it to. If we shake the bag to make the marbles move faster, what will happen to the bag?

> **Charles's law** states that for a mass of gas at a constant pressure, the volume and temperature are directly proportional:
>
> $$\frac{V_1}{T_1} = \frac{V_2}{T_2}$$

> **WORKSHEET**
>
> Download this unit's worksheet and do activities 21 and 22.

An **increase in temperature** will make gas particles move faster. Using a piston (a lid that can move up and down), we can maintain a constant pressure. This gives particles freedom to move. When we increase the temperature, the particles move faster and occupy more space. This causes an **increase in volume**. We must express the temperature in Kelvin.

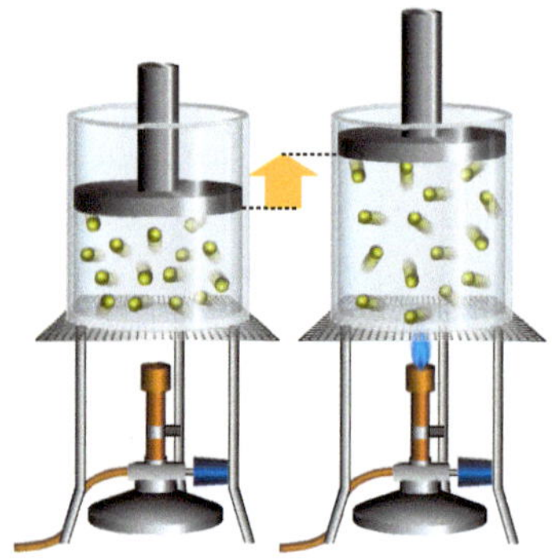

EXPERIMENT Generating space

9. Inflate a balloon and put it between two objects (two books or two cartons of milk, for instance). The balloon should fit between the two objects without pressing it, as shown in the image. Later, put the balloon in the freezer for 30 minutes. Take it out and put it between the objects again.
   a) Is the space around the balloon the same now?
   b) Explain what has happened using the kinetic theory.
   c) Prepare a short report of your experiment, indicating the procedure and the results. Record the temperature in the room and in the freezer, as well as the separation between the objects.

ACTIVITIES

22. **We have a cylinder with a moveable piston. We fill it with 25 cm³ of air at a temperature of 7 °C. The cylinder's maximum volume is 30 cm³. What is the maximum temperature we can heat the cylinder to at a constant pressure? Always express the temperature in the SI unit.**

23. **Hot air balloons are an example of the applications of Charles's Law. Explain what happens to the balloon when we heat it. Use the kinetic theory. Discuss your answer with a partner and find a common explanation.**

5.3. Gay-Lussac's law

Joseph Louis Gay-Lussac described this law at the beginning of the nineteenth century. The French chemist studied the relationship between pressure and temperature. To do this, he kept the volume constant.

THINK AND REASON

Once again, we have some marbles in a bag. However, this time the bag cannot change its size. If we shake the bag continually, will the marbles hit the sides of the bag more frequently?

Gay-Lussac's law states that for a mass of gas at a constant volume, the pressure and temperature are directly proportional:

$$\frac{P_1}{T_1} = \frac{P_2}{T_2}$$

In this case, the piston does not move. Because the piston cannot move up or down, the volume stays constant. When the **temperature increases**, the particles move faster. Because the volume remains constant, there are more collisions against the walls. In other words, the **pressure increases**. We must express the temperature in Kelvin.

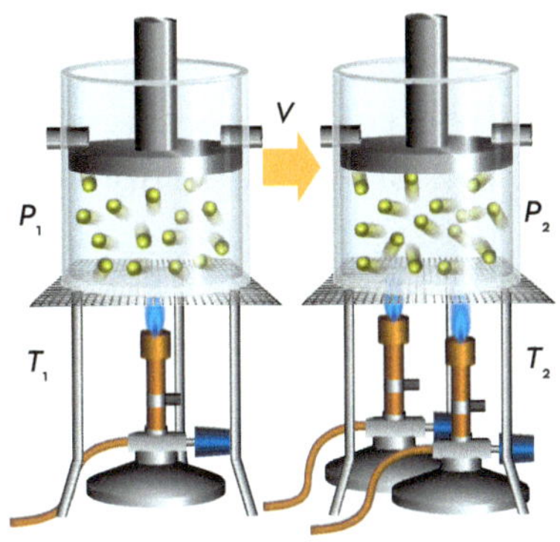

DID YOU KNOW?

A pressure cooker is an example of the applications of Gay-Lussac's Law. The volume remains constant so an increase in temperature results in an increase in pressure.

WORKSHEET

Download this unit's worksheet and do activity 23.

SOLVED EXAMPLE

5. We put gas in a closed receptacle at a pressure of 1 atm and heat it from 20 °C to 390 K. We keep the volume constant. Calculate the value of the final pressure.

Solution

The volume remains constant so we have to apply Gay-Lussac's Law. We must convert the initial temperature to Kelvin (the SI unit).

$P_1 = 1$ atm $\quad T_1 = 20 + 273 = 293$ K $\quad P_2 = ?$ $\quad T_2 = 390$ K

$$\frac{P_1}{T_1} = \frac{P_2}{T_2} \rightarrow \frac{1}{293\text{ K}} = \frac{P_2}{390\text{ K}} \rightarrow P_2 = 1.33\text{ atm}$$

ACTIVITIES

24. Keeping a fixed volume, we increase the pressure of a gas from 2 atm to 5.2 atm. Calculate the initial temperature of the gas if the final temperature is 53 °C.

25. We heat a receptacle with a constant volume. The initial temperature is 30 °C and the final temperature is 50 °C. The initial pressure is 2 atm. Calculate the final pressure.

MIND MAP

Download and print the worksheet you will find in the OLC to complete the mind map of the unit.

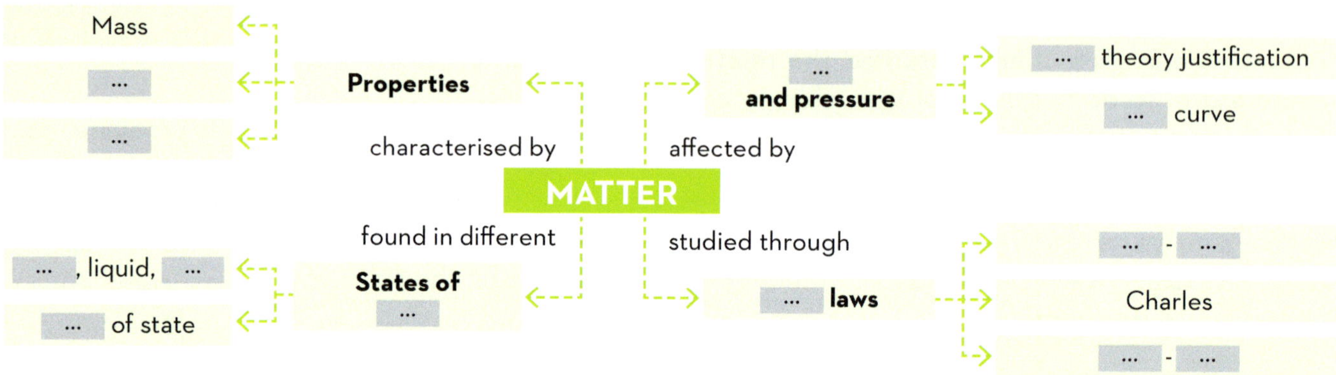

LOOK AROUND YOU

Coincidence and science. Eureka!

Hiero was the king of Syracuse in the third century BC and he was related to Archimedes. One day Hiero gave Archimedes a problem to solve. A goldsmith had made the king a gold crown but Hiero was not sure if the artisan had been honest. The king thought the goldsmith had kept some of the gold for himself and substituted it with silver or copper. Hiero asked Archimedes to find out if the crown was pure gold without destroying it.

Archimedes did not know what to do. Copper and silver are lighter than gold. If he could calculate the space the crown occupied (the volume), he could answer Hiero's question. But he did not know how to calculate the volume of the crown without transforming it into a compact mass.

Archimedes was considering the problem in the public baths, probably sighing in frustration. As he submerged himself in a full tub, he observed how the water spilled over the sides. He suddenly leaped to his feet: he realised that his body displaced the water out of the bath. The volume of the displaced water had to be equal to the volume of his body. He only had to measure the volume of displaced water to calculate the volume of any object. In a flash of inspiration he had discovered the principle of displacement!

Archimedes could not wait to test his theory. He jumped out of the bath and ran through the streets naked and soaking wet, shouting, 'I have found it, I have found it!' Only he shouted in Greek of course: 'Eureka! Eureka!'

At home, he filled a receptacle with water, put the crown in it and measured the volume of the displaced water. Then he did the same with another object with the same weight and made of pure gold. He proved that the goldsmith had made the crown using gold mixed with a lighter metal. The king ordered the execution of the artisan.

Adapted from *Breakthroughs in Science*. Isaac Asimov.

Project

Working with a partner, write five questions that can be used to interview Archimedes. Brainstorm with the rest of the class to select 5 to 10 questions suitable for an interview. Then, working with your partner again, do an interview in which one of you is a journalist and the other one is Archimedes. Record your interview and transcript and share with the rest of the class.

LABORATORY WORK

Determining densities

Objective

In this experiment, we are going to calculate the density of three different objects: marble, iron and alcohol. We will prove that density is a specific property and understand why it is useful.

Material

- Beaker.
- Crucible.
- Graduated cylinder.
- 96% alcohol.
- Block of marble.
- Scales.
- Iron filings.

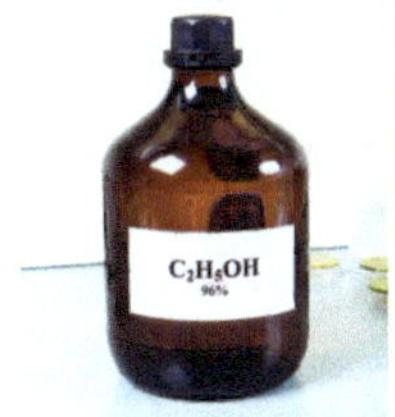

Procedure

- To calculate the **density of the two solids**, we use the following procedure:
  1. Determine the mass. To do this, put the crucible on the scales. Adjust the scales to 0 grams (called tarring the scales). Put the solid in the crucible. Write down the value (*m*).
  2. To determine the volume it occupies, put water into the graduated cylinder (approximately half full). Write down the value (V_0).
  3. Put the solid in the cylinder and write down the new volume (V_1).
  4. The difference between V_1 - V_0 is the volume the solid occupies (*V*).
  5. Follow this procedure first with the iron filings and then with the block of marble.
- To calculate the **density of the alcohol**, we use the following procedure:
  1. Put some alcohol in the graduated cylinder (approximately half full). Write down the value (*V*).
  2. Put a beaker on the scale and tare it.
  3. Slowly pour the alcohol into the beaker. Write down the value (*m*).

Task

1. In your notebook, write all the data into a table like this one.
2. Search for the real density value of the three substances and compare them with the results you have obtained in the laboratory. Comment on the results. If there are any mistakes, why do you think there are?
3. Is it possible to calculate the density of cork with this procedure? Explain your answer.

	Marble	Iron	Alcohol
Mass	...	...	...
V_1	...	...	...
V_0	...	...	...
$V = V_1 - V_0$	...	...	...
$d = (m/V)$	...	...	...

FINAL ACTIVITIES

Basic activities

1. Say whether the following statements are true or false and explain your answer. If they are false, rewrite them so that they are true.

a) The density of solids is greater than the density of liquids.

b) If the density of a solid is greater than that of water, the solid will sink in this liquid.

c) Density depends on how much of the substance we have.

2. Using a graduated cylinder, we measure 200 mL of olive oil and pour it into a glass. We weigh the glass and its contents, obtaining a value of 456 g. The mass of the glass is 272 g. What is the density of the oil? Express the result in g/cm³. Compare your answer with a partner.

3. Which of the following statements is true? Explain your answer. Change the false statements so that they become true.

a) Vaporisation is the transformation of a solid to a gas.

b) Melting is the transformation of a solid to a liquid.

c) Condensation is the transformation of a liquid to a gas.

4. Calculate the volume of 2 kg of polystyrene in the SI unit. Density = 920 kg/m³.

5. We take a bottle of fizzy drink out of the fridge. After a short time, we observe the bottle is wet on the outside. Where does this water come from? What has happened?

6. Look at the diagram below. Which state of aggregation does it represent? Indicate the characteristics of this state.

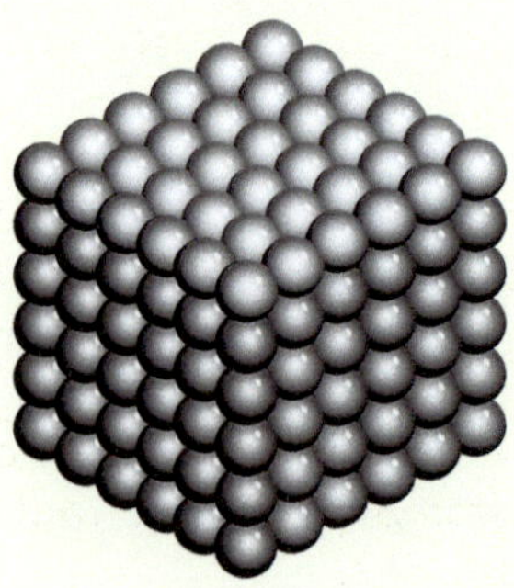

7. We have a gas at a constant temperature with an initial volume of 500 mL at a pressure of 2.3 atm. We reduce the pressure to 1.7 atm. Calculate the new volume of the gas.

8. The melting point of aluminium is 660 °C and its boiling point is 2 519 °C.

a) What state is the aluminium in at a temperature of 800 °C?

b) What is the condensation point of aluminium?

9. We have a gas in an enclosed receptacle with a constant volume. We have obtained the following values.

T (K)	260	390	536	627	660
P (atm)	0.80	1.20	1.65	1.93	2.04

a) Draw these values on a graph. Put *T* on the horizontal axis.

b) Indicate which law this graph represents.

c) Predict the temperature the gas will reach at 5 atm of pressure.

10. Do the following conversions:

a) 100 g to kg.
b) 2 atm to Pa.
c) 3 dm³ to cm³.
d) 68 ºC to K.
e) 5 m³ to dm³.
f) 2.45 kg to cg.

11. Use the kinetic theory to explain how a glass of water evaporates.

12. Calculate the volume of a basketball with a diameter of 24 cm.

13. Calculate the density of an object. To do this, we measure its mass on some scales, obtaining a value of 50.2 g. We also measure its volume in a graduated cylinder. The water initially has a volume of 22 cm³. When we put the object in the water, the volume rises to 34 cm³. Write the procedure you follow to solve this problem.

14. We have a mixture of petrol, water, oil and alcohol. Indicate in which order the liquids will separate into layers because of their respective densities. To do this, search for their density values. Compare your answer with a partner.

Consolidation activities

15. During an excavation, archaeologists have found a metal object. They want to know if it is copper or a different metal. They put it on some scales and obtain a value of 137 g. When they submerge it in water, it displaces a volume of 15.4 cm^3. Data: density of copper is 8.93 g/cm^3.

a) Find out if it is copper.

b) Can you use another property to be sure?

16. Put the following substances in order of density from highest to lowest: alcohol (0.8 g/cm^3), water (1 000 kg/m^3), silver (0.0105 kg/cm^3) and oil (0.9 g/cm^3).

17. You have some alcohol in a glass. You pour half of the alcohol onto the table. After a certain time, the table is dry, but the glass still contains alcohol. Use the kinetic theory to explain this phenomenon.

18. Write a short story in which water molecules are the characters to explain how ice melts into water.

19. Create a Venn diagram to show the differences and similarities between the liquid state and the gaseous state according to the kinetic theory.

20. Pollutant gases have a more damaging effect when they are in open air. Work with a partner to explain this fact.

21. A specific mass of gas is contained in a rigid receptacle. At a temperature of 573 K, it has a pressure of 2 atm. What pressure will the same amount of gas reach if we reduce the temperature to 473 K? Which gas law did you use?

22. Working at a constant pressure, we have a gas at a temperature of 34 °C occupying a volume of 300 mL. If we halve the volume, what is the resulting temperature?

23. Imagine a glass of water. Draw a picture showing the distribution of particles in the water.

Advanced activities

24. Complete the following table in your notebook:

Matter	Mass (kg)	Volume (m^3)	Density (kg/m^3)
Paper	...	0.00167	750
Air	60	...	1.29
Diesel	3 500	5	...

25. Look at the heating curve for ethyl alcohol:

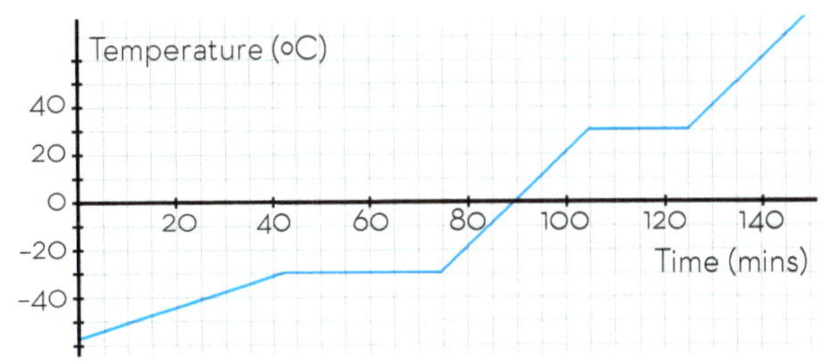

a) Mark the melting point and the boiling point on the graph.

b) Mark on the graph the temperature range where it is in a liquid state.

c) What state is the ethyl alcohol in at a temperature of –20 °C?

d) Compare your answers with a partner.

26. We find most substances in the three states of aggregation we have mentioned. However, there are other states of matter. Working in threes, create a presentation indicating the name, characteristics and uses of all the states of aggregation known to us.

27. Mechanics always tell us it is better to measure the pressure in our car tyres when the wheels are cold. (In other words, driving as little as possible beforehand.) Investigate why this is. You will need the kinetic theory. Prepare a short talk to explain this fact and share with your classmates.

USE YOUR SKILLS

Wind power

Many people think that wind power is an energy source that can replace oil or coal power stations. The structures in the photo below are wind turbines with blades turned by the wind. These revolutions produce electricity in generators moved by the rotary blades.

Question 1

The graphs below show the average wind speed in four different locations throughout the year. Which graph indicates the most suitable location for the installation of a wind turbine?

Compare your answer with a partner and prepare a common explanation.

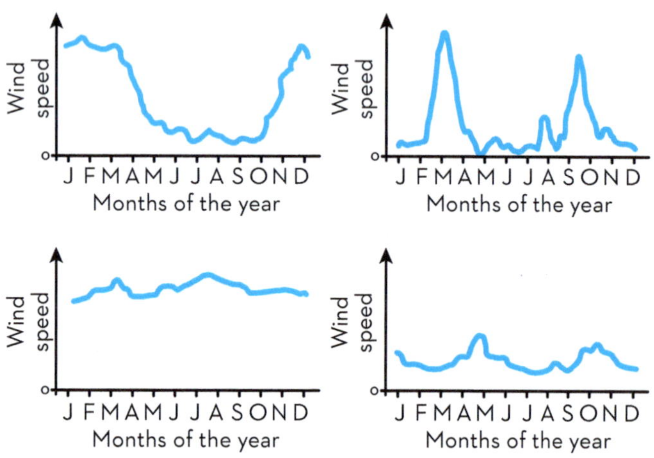

Question 2

Wind turbines located at higher altitudes revolve more slowly, even if the wind speed is the same as at lower altitudes.

Which of the following reasons best explains why the blades turn more slowly at higher altitudes?

a) The air is less dense at higher altitudes.

b) The higher the altitude, the lower the temperature.

c) The higher the altitude, the less gravity there is.

d) It rains more often at higher altitudes.

Question 3

Give one advantage and one disadvantage of the production of electricity through wind power compared with the production of electricity through fossil fuels such as coal and oil.

Question 4

There are onshore and offshore wind farms. Offshore wind farms are located in the sea, not on the land.

e) What are the characteristics of the sea winds?

f) Give an advantage and a disadvantage of offshore wind farms.

The mysterious bottle

→| Competency task

Objective

This task involves using your newly acquired knowledge to help our agents solve a mysterious case.

Our team of investigators are very aware of environmental problems. In one of their daily visits to the river, they have found a plastic bottle with no label on the riverbank. Someone has written 'highly flammable organic solvent' with a marker on the bottle.

We need to find out what the bottle contains so they can take it to the appropriate storage location. The investigators quickly took the bottle to the laboratory so the technicians could do the relevant tests. To do these tests, they took a sample of the liquid, heated it and drew the following heating curve based on their results.

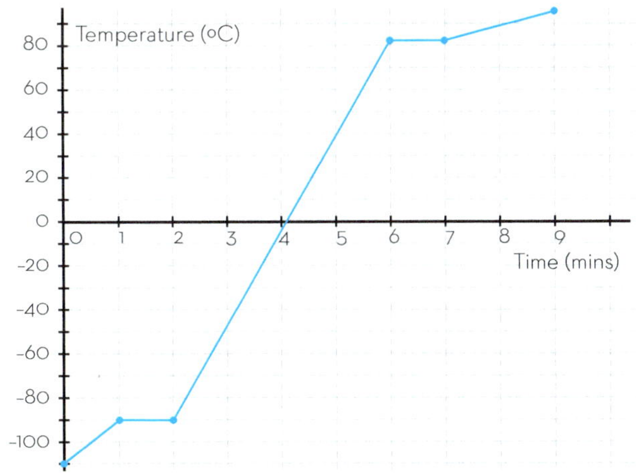

To resolve this case, the laboratory has a table with information on the most commonly used alcohols for solvents:

Common Name	mp (ºC)	bp (ºC)
Methyl alcohol	−97.8	65.0
Ethyl alcohol	−114.7	78.5
Propyl alcohol	−126.5	97.4
Isopropanol	−89.5	82.4
Isopropyl chloride	−117.2	35.7
Propane	−187.7	−42.1
Butyl alcohol	−89.5	117.3
Tert-Butyl alcohol	25.5	82.2
Amyl alcohol	−79	138
Neopentyl alcohol	53	114

Variability of physical properties in alcohols.

mp = melting point
bp = boiling point

Final product

Your final product will be a presentation with the following elements.

1. Explain which liquid is in the bottle and the procedure you followed to find out, relating this procedure to the scientific method.
2. List the specific characteristics of the liquid and what hazard diagrams should have been shown on the bottle.
3. A report on the consequences of pouring this type of pollutants in fresh water sources, such as rivers or lakes.
4. Find information about the characteristics of the alcohol in the bottle. Draw the hazard diagrams the bottle should have on it. Indicate what the correct response to these hazards should be.
5. Search for information about the risks and consequences of this type of water pollution and the adequate procedures to deal with these waste chemicals.

Self-assessment

Evaluate your work and your classmates' work with the guidelines that you can download from the OLC.

1st STAGE: IDENTIFYING THE PROBLEM

! IMPORTANT

To do this research, it is useful to know that the **density of water** is:

1 kg/L.

In this unit you have studied density. Can you apply your knowledge to a research project?

We are going to have some fun using a simulator now. You can find it at **goo.gl/Qyg7pt**. Your mission is to discover which mystery objects are hiding at the end of the simulation and answer the question:

WHAT CHARACTERISTICS MUST AN OBJECT HAVE TO FLOAT IN WATER?

2nd STAGE: FORMULATING THE HYPOTHESIS

There are many possible hypotheses for this subject. However, use your current knowledge to formulate some hypotheses (possible answers) to these specific questions:

[H1] What density must an object have for it to float?

[H2] Imagine we have two objects with the **same mass** and only one of them floats. Which one floats? The large one or the small one?

[H3] Now we have two objects with the **same volume** and only one of them floats. Which one floats? The heavier one or the lighter one?

3rd STAGE: RESEARCH

Experiment 1: Making contact

First you should familiarise yourself with the simulator.

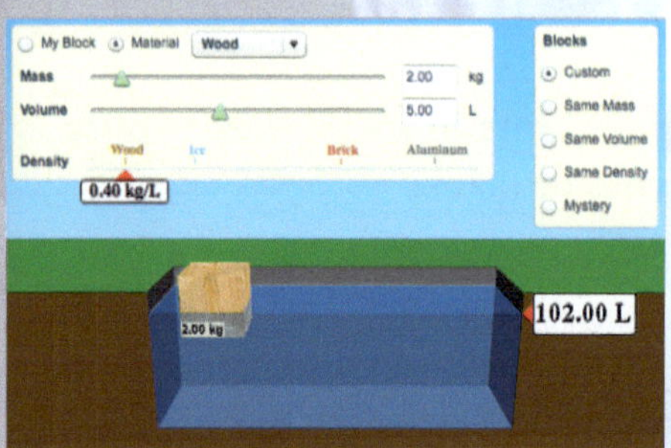

1. Play with the **wooden** object. Sink it, drop it, submerge it, do anything you like with it. Then let it rest, follow these instructions and do the calculations:
   a) Leave the mass at 2 kg, read its volume and check its density.
   b) How can you find out the submerged volume?
   c) If the object's volume were only the submerged part, what would its density be?
   d) Increase its mass to 4 kg and answer all the questions again.

5. Change the object to **ice** and repeat the experiment.
6. Now do it one last time with **aluminium**. What happens?
7. Select the *Custom* option on the right. Now select the *My Block* option on the left. Change the mass and/or volume to make objects float or sink.

Verify your first hypothesis (H1). Rewrite it if necessary.

Experiment 2: Objects with the same mass

Select the *Same Mass* option on the right. Determine the following information and complete a table like the one shown here. Show your work to find out:

a) The volume of each object. *b)* The density of each object.

Verify your second hypothesis (H2). If it was not correct, what should it be?

Object	Mass	Volume	Density	Floats
Blue	...	...	...	...
Yellow	...	...	...	...
Green	...	...	...	...
Red	...	...	...	...

Use this table as a reference for all the others.

Experiment 3: Objects with the same volume

Now select the *Same Volume* option, answer the questions and complete a new table.

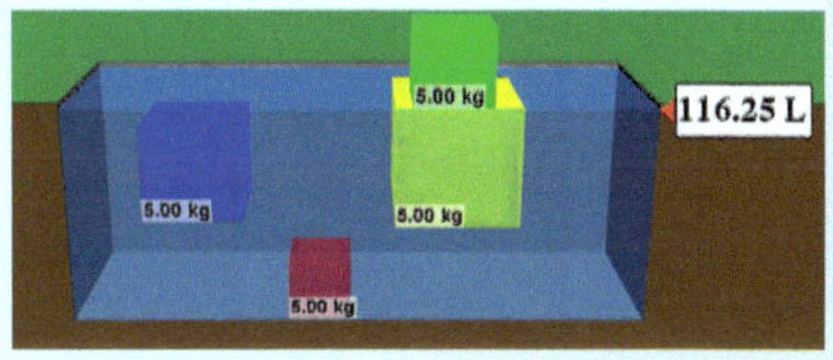

h) What is the volume of each object?

i) What is the density of each object?

Verify your third hypothesis (H3). Rewrite it if necessary.

Experiment 4: Objects with the same density

Select *Same Density* and answer the following questions:

a) Determine the mass and volume of each object and complete a third table.

b) Calculate what the density of each object is if they only occupy the submerged volume.

c) Take the 4 kg object and put it on top of the 3 kg object. What happens? Now answer question *b)* again.

Experiment 5: Mystery objects

This case is simple. Using your knowledge, you must find out the density of all the objects that appear. When you have finished, select *Show Table* to discover which material they are all made of.

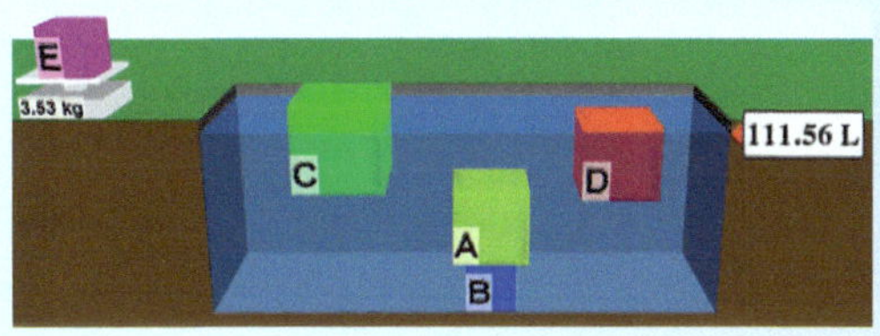

4th STAGE: MATERIALS AND METHODS

Put all your tables together (four in total) and take some screenshots of the simulator. The screenshots should show at least one case for each experiment. Write brief summaries of what happened in each case.

5th STAGE: CONCLUSION

Using all the information and material you have collected as well as your own personal opinion, create a presentation to project or watch on a screen. If you do not have a projector or screen, use a piece of cardboard for each experiment and stick your screenshots on them. Prepare a poster with your cardboard pieces. Write a brief text explaining what happens to describe each case. Finally, give an oral presentation of your entire investigation.

IMPORTANT !

Remember, you can do more tests by selecting the *Custom* option on the right and then the *My Block* option on the left.

3 MATTER SYSTEMS

People can be grouped in different ways. Imagine a group of football players that are part of a team, all with the same shirt colour. When we first look at them, we cannot distinguish one from the others. But if we watch a football match, we will observe that the players from one team meet with those from the other team on the field and we can distinguish shirts of different colours mixed together.

This also happens with matter. Different substances can be put together to obtain mixtures of them. There are mixtures in which we can observe different particles together, like the football players in the match, and there are mixtures in which all the particles seem indistinguishable from the others, like the football players on the same team. These mixtures are called solutions.

Contents

WORK WITH THE TEXT

The text shows an analogy or metaphor to explain an idea. In your notebook, match each term in the first column with its simile in the second.

1. Football player.	a. Solution.
2. Football team.	b. Mixture.
3. Shirt colour.	c. Mixing.
4. Football match.	d. Observable property.
5. Grouping.	e. Particle.

CHALLENGE

Put some sugar, sawdust and sand in a bowl. Mix these substances completely.

- Can you differentiate the components of the mixture?
- Can you figure out how to separate them now?
- Put some water into the bowl. Is it easier to separate any component now?
- What happened to the sand and sawdust?
- Where is the sugar? (Hint: Take a coffee filter and pass the water through it. Taste a drop of the water.)
- Is there any way that we can get the sugar back?

Write a short report indicating the tasks that you performed to separate each component, your observations and final results.

Have a look at a mineral water bottle label. Check the composition.

- How many components can you read?
- Can you see them inside the bottle? Why?

Take another bottle of a different brand. Does it have the same composition?

KEYWORDS

Download and print the activities that will guide you through the essential vocabulary of this unit.

FIND A CLASSMATE

Download and print the worksheet 'Find a classmate: a classification of food'. Share experiences about this topic with your classmates and discover interesting things about them.

1 Classifying matter

Everything that we are and everything that surrounds us is matter. Matter occupies space and has mass. The air we breathe, the chair we sit on, the clothes we wear, the food we eat, the water we drink, our own body, everything is matter.

Matter is everything that has **mass** and occupies **space**.

EXPERIMENT

1. Matter is there even though you cannot see it.
   - Submerge an 'empty' tube upside down in a receptacle with water.

   Describe what you observe. Why doesn't the water enter the tube?
   - Now measure the mass of an uninflated balloon and then with the balloon inflated.

   Do they have the same mass? Did the mass vary?

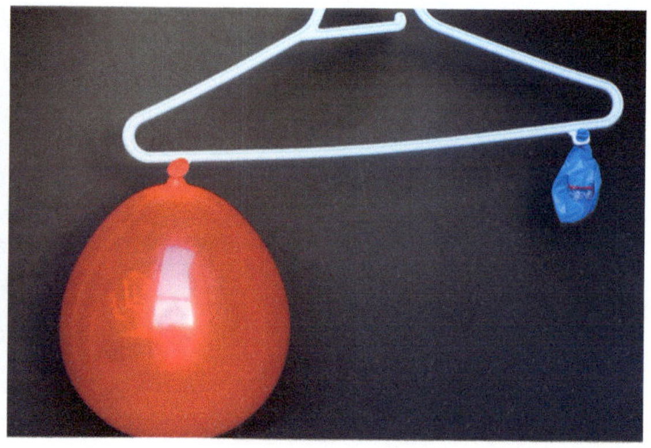

We said that matter has mass and occupies space. This means that we can measure it.

- The amount of matter of an object comes from its **mass**. This is normally measured in kilograms or in multiple or submultiple of this unit.
- The **volume** of an object is measured by the space it occupies, usually in litres or in cm^3.

And like everything, matter can be classified according to its characteristics and behaviour. These are the specific properties it has and which differentiate it from other types of matter:

- **Physical properties:** These depend on the substance itself. For example, colour, smell, texture, flavour, density, etc.
- **Chemical properties:** These depend on the interaction of matter with other substances. For example, the oxidation of an iron nail is caused by oxygen in the air.

SCIENCE 2.0

Watch the video at:

goo.gl/uQDgxW

Explain matter and its properties.

WORKSHEET

Download this unit's worksheet and do activities 1 and 2.

Physical properties can be classified into two groups:

- **Extensive physical properties:** These depend on the amount of matter. They are mass, volume and length.
- **Intensive physical properties:** These depend only on the type of matter. For example, density, colour, hardness, etc. They do not depend on the amount of the matter: a litre of water has the same density, the same colour and the same hardness as five litres of the same water.

EXPERIMENT

2. Prepare several receptacles with the following matter systems:

water

sand

salt

oil

sugar

Mix all of them one by one and indicate what happens in each case. Prepare a table like the one on the side. It will help you analyse the results more easily.

Mixture	Result
Water and sand	...
Water and salt	...
Water and oil	...
Water and sugar	...
Sand and salt	...
Sand and oil	...
Sand and sugar	...
Salt and oil	...
Salt and sugar	...
Oil and sugar	...

Matter systems can be classified as:

- **Pure substances:** Whatever their states of aggregation, their physical and chemical properties are uniform in any part of them. These are characteristic of them and differentiate them from other substances.
- **Mixtures:** These are two or more pure substances combined together. The proportion may be variable.

ACTIVITIES

1. **Indicate six matter systems and classify them as pure substances or mixtures. Compare your classification with a partner to create a twelve items list.**
2. **Identify which ones from the following are matter: the lightning bolt of a storm, the sound of a bell, the light from a light bulb and rain.**

WORKSHEET

Download this unit's worksheet and do activity 3.

1.1. Mixtures

Most of the time, matter around us is in the form of mixtures. We rarely find pure substances. Depending on the particle size of the substances, mixtures may be **homogenous colloidal** or **heterogeneous** and their components can be separated by **physical methods** or **mechanical methods.**

THINK AND REASON

Find the following substances:

- Milk.
- Wine.
- Coffee.
- Paella.
- Tea.
- Mayonnaise.
- Liquid yogurt.
- Freshly-squeezed orange juice.
- Cream of courgette.

a) Examine each one of them and decide if they are made up of one or various substances. Write down your observations.

b) Now look for information on the internet about each one of them, and indicate if your previous observations were correct.

Heterogeneous

Heterogeneous mixtures are substances that are made up of two or more components. We can easily distinguish these components just by looking at them. Their particles are larger than 10^{-6} m.

Their properties and composition vary from one part to another of the same sample.

The components of heterogeneous systems can be separated by mechanical methods (filtration, decantation, etc.)

Gravel and water

Oil and water

Characteristics:

- They are filterable.
- They separate spontaneously by gravity.

Colloidal systems

Between true heterogeneous systems and homogeneous systems there are **colloidal systems.** These are mixtures that have at least two phases: one is finely divided into small particles (dispersed phase or micelle) and is surrounded by the other substance (dispersion medium or continuous phase).

But what does finely divided mean? This means they have properties that are clearly colloidal when the dimensions of the dispersed phase are between one millionth (10^{-6} m) and one billionth (10^{-9} m) part of a metre. The dispersed phase can often be seen with a microscope.

$< 10^{-6}$ m

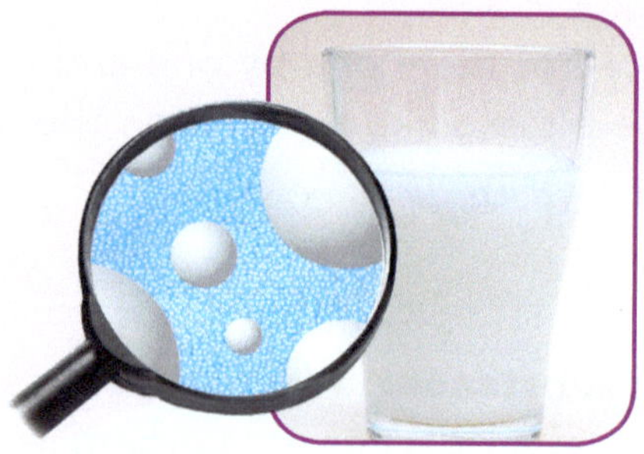

Colloid and dispersed phase

The following are colloids:

All living tissues like blood.

Many foods, like cheese, butter, clear soups, jellies, mayonnaise, whipped cream and milk.

Numerous products that we use today: rubbers, plastics, paints, lacquer, varnish, among others.

The main types are indicated in the following table:

Dispersed phase → / Continuous phase ↓	Solid	Liquid	Gas
Solid	---	gels *jelly*	foams *pumice stone*
Liquid	sols *paint*	emulsions *mayonnaise*	foams *whipped cream*
Gas	---	aerosol *spray*	---

Type of system
Example

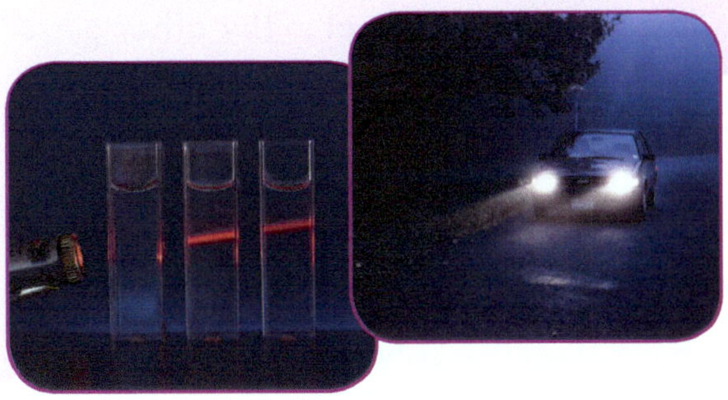

Colloids are easily distinguished from homogeneous systems by the **Tyndall effect**. This is the dispersion of light when it passes through colloids, allowing the light to be seen through colloids. An example is the dispersion of light through dust particles in the beam of a torch.

Water and salt

$< 10^{-9}$ m

Homogeneous

Homogeneous mixtures are substances that are made up of two or more components. The size of the particles is less than 10^{-9} (one billionth of a metre). They cannot be distinguished visually.

Their properties and composition are uniform in any part of a given sample.

The components of homogeneous systems are separated by methods that involve changes of state, like evaporation, recrystallisation, distillation, etc.

ACTIVITIES

3. **Identify if these are homogeneous or heterogeneous mixtures: air, salt water, concrete, bronze, granite.**

4. **Marketing language talks about 'pure cow's milk', 'pure spring water', for example. Indicate if these statements are true. Look at the labels of some of the brands you use at home and explain your answer. Discuss the meaning of 'pure' with a partner.**

5. **In groups of four, make a poster with examples of the main types of colloids in everyday life. Then explain the poster to your classmates.**

SCIENCE 2.0

Learn about the Tyndall effect at:

goo.gl/RTy4GN

Reproduce the experiment at home.

2 Solutions

We can find numerous substances that are not pure, like the air we breathe, sea water, vinegar, etc.

A **solution** is a homogenous mixture of two or more components.

EXPERIMENT Floating alcohol

3. Prepare two transparent glasses and the following substances: water, oil and alcohol.

Pour some water in the first glass and add oil. Then add alcohol. Stir the three components in the glass with a spoon and then let them set.

1. Describe what happens in each of the previous phases.
2. What conclusion can you make from this?
3. What would happen if you mix just water and alcohol in the other glass?

When a substance dissolves in another, the particles of one component separate and then disperse in the other.

- The component in smaller proportion is called the **solute.**
- The component in larger proportion is called the **solvent.**

WORKSHEET

Download this unit's worksheet and do activities 4 and 5.

IMPORTANT

When the components of a solution are in similar proportions, the solvent is the substance that is in the same state of aggregation as the final solution.

ACTIVITIES

6. Complete the table in your notebook.

Solution	State of solution	State of solute	State of solvent
Sugar water	...	...	...
Surgical spirit	...	...	...
Champagne	...	...	...
Air	...	...	...
Bronze	...	...	...

2.1. How solutions are formed

The particles of the solute and the particles of the solvent mix together to form the solution. But how does this process occur at a molecular level? Most importantly, how can we make this happen more quickly?

EXPERIMENT

4. Take a glass and fill it with water. Add some drops of ink with a dropper (or something similar) so that they drop slowly.

   Describe what you observe. Wait for an hour and observe it again. What is happening?

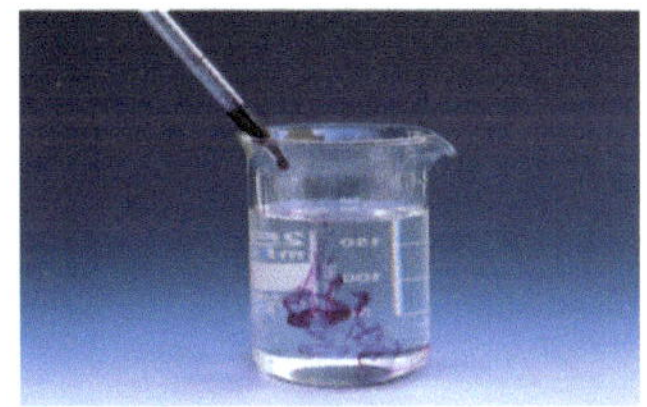

The **kinetic theory of matter** that you studied in the previous unit helps explain this phenomenon:

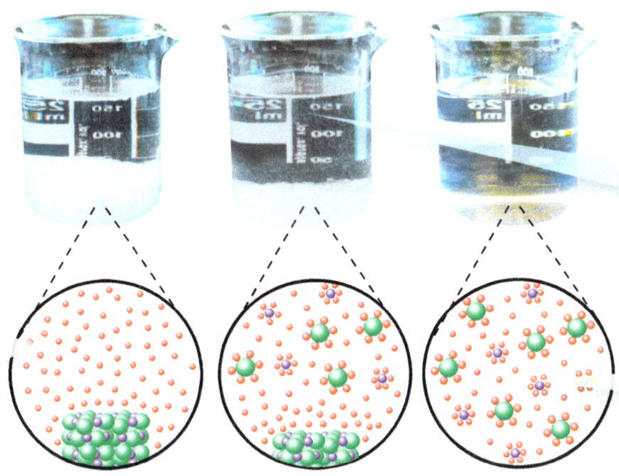

It is easy to imagine how we can influence the formation of the solution:

What happens if we stir the components with a spoon? What happens if we increase the temperature of the receptacle containing them?

In both cases the particles will receive more energy. This means that their speed will increase and their bonds will weaken. This will facilitate dispersion, thus obtaining a solution.

WORKSHEET

Download this unit's worksheet and do activity 6.

ACTIVITIES

7. **On more than one occasion you have seen how you put a spoonful of sugar in water and it dissolves little by little. If you stir it, you observe how the sugar crystals dissolve more quickly.**

   Make a drawing with the sugar molecules and water molecules before dissolving. With help from the kinetic theory of matter, draw how the process of dissolution occurs. What will be the mass of the solution after dissolution? Write the procedure of a simple experiment to determine changes in the mass of the solution.

SCIENCE 2.0

Look at this animation which shows how a sugar cube dissolves in water at a molecular level:

goo.gl/qegckX

Some examples of commercial solutions.

3 Types of solutions

Solutions can be of different types depending on the physical state of the components or on the relative amounts of them.

3.1. Depending on the physical state of the components

As we already know, matter can be present in three states, solid, liquid and gas. The possible combinations of these will give us different types of solutions.

THINK AND REASON

Imagine you are in three locations: in the street, in the kitchen of your home and in a cafe. Answer the following questions:

a) Can you identify any solution?

b) Can you identify its components? Do you know which one is the solute and which one is the solvent?

We can see all the possible combinations of solution systems organised in the following table:

Solvent	Solute	Examples
Solid	Solid	Alloys: bronze, brass, steel
	Liquid	Amalgams
	Gas	Activated charcoal tablets (absorb gases)
Liquid	Solid	Sea water
	Liquid	Isopropyl alcohol
	Gas	Carbonated drinks, household ammonia
Gas	Solid	Smoke
	Liquid	Fog
	Gas	Air

WORKSHEET

Download this unit's worksheet and do activity 7.

IMPORTANT

Some of the solutions proposed (fog and smoke) can also be colloidal systems if the molecules of the solute group together to form particles of a sufficient size (between 10^{-6} and 10^{-9} m).

ACTIVITIES

8. Identify the components of the examples included in the table above by looking up information in books or on the internet: bronze, brass, steel, amalgams, sea water, isopropyl alcohol, carbonated drinks, smoke, fog and air.

Create a table with the headings Solution, Solvent and Solute and indicate the components for each mixture.

Gaseous solutions

As the particles of gases move freely, they spread out easily, obtaining a homogeneous mixture very quickly.

The most important example is air. It contains a mixture of gases in slightly variable proportions: **nitrogen** (78 %), **oxygen** (21 %), and others like **carbon dioxide** (0.03 %) and **argon** (0.93 %).

SCIENCE 2.0

This video explains how air is essential for life:

goo.gl/OJPNO5

EXPERIMENT **Playing with the baby bottle**

5. Take a baby bottle and glue the hole of the teat closed. Pour a fizzy drink into the bottle and close it. The drink should be rather cold and you must open it carefully, without shaking it, and pour it into the bottle quickly. Then heat the bottle with your hands and shake the teat.

   What happens at the beginning? And after heating it? Explain.

   Is the pressure of the gas escaping very strong? Try inflating the teat before the experiment by blowing in it.

Liquid solutions

The most important **liquid solutions** are those that involve **water**.

When the forces with which the solvent molecules attract the particles of the solid are greater than the forces acting between them, the particles separate and scatter forming a solution.

For liquids to form a solution they must have the property of being completely miscible, like water and alcohol, for example, and not water and oil.

Gases always dissolve in water, in larger or smaller proportion, depending on the pressure of the gas. This is how they make carbonated drinks like cola, tonic water, etc.

WORKSHEET

Download this unit's worksheet and do activities 8 and 9.

Solid solutions

The most important ones are **alloys**. An alloy is a mixture of metals that is obtained by mixing together the melted components and cooling the mixture. In these solid solutions, the atoms of one metal randomly distribute among the atoms of the other metal, although in some cases defined structures are obtained. Some examples are bronze, steel and brass.

ACTIVITIES

9. As a group, brainstorm how the miscibility of gases facilitates environmental pollution processes.

10. Why do you think that the bubbles of a fizzy drink escape so easily when we open the bottle? Share your ideas with a partner and write a short explanation of this fact.

IMPORTANT

The oxygen gas dissolved in the water of rivers and seas allows animals and plants to live underwater.

3.2. Depending on the proportion of the components

SCIENCE 2.0

This applet shows how salt is dissolved and how saturation is reached:

goo.gl/Q1q2C6

The amount of a solute that can be added to a solution is variable. There are different types of solutions depending on the relative proportions of the components.

Solubility is the maximum amount of a solute that can be dissolved in a certain amount of solvent.

This is a fixed proportion for each pair of solute and solvent at a certain temperature.

Solubility can be expressed in grams of solute per litre of solvent (g/L).

EXPERIMENT

6. Take three glasses with 100 mL of water and copper sulphate. Add a few crystals to the first glass. Add a spoonful to the second and three spoonfuls to the third. Stir until it is all dissolved.

   Describe what you observe in each glass.

   Do you think you could dissolve all the copper sulphate you want? Explain.

   What happens if you increase the volume of water in each glass?

WORKSHEET

Download this unit's worksheet and do activities 10 and 11.

As you can see, there are different types of solutions depending on the amount of solute that is dissolved:

- **Dilute solution:** the amount of solute is small in comparison with the solubility.
- **Concentrated solution:** the amount of solute is near solubility.
- **Saturated solution:** the solution will not admit more solute without adding more solvent at a certain temperature. The proportion between the solute and the solvent is the highest possible. Some solute appears at the bottom of the receptacle because it cannot dissolve.

Depending on the characteristics of each solute and each solvent, saturation will be reached sooner or later.

THINK AND REASON

Imagine you have a saturated solution.

How do you think you could add more solute and it would continue dissolving?

What do you think would happen if you heat the solution and add more solute? Check this by experimenting.

- A **supersaturated solution** is a solution that has a greater amount of solute dissolved in it than it can dissolve. This is achieved by increasing the **pressure** (for gas solutes) or the **temperature** (for solid solutes) and then letting it cool slowly so that when it cools the excess does not precipitate to the bottom of the glass.

There are some factors that affect solubility:

- **Contact surface:** when this increases, favoured by the pulverisation of the solute, the **speed** of dissolution increases because the interactions between the particles of the solute and the solvent are greater.
- **Stirring:** when the solute is dissolved, the particles must spread throughout the solvent. Stirring favours **mobility**, increasing the dissolving rate.
- **Temperature:** this affects the **speed** and the **degree of solubility**. Increasing the temperature favours the movement of the molecules in the solution. It also causes the **energy** of the solute particles to increase, which causes them to disperse more easily. For gases this process of increasing the speed by increasing the heat causes just the opposite effect: it decreases their solubility.
- **Pressure:** for gas solutes, their **solubility** is directly proportional to the pressure. It is easy to observe how opening a bottle of fizzy drink decreases its pressure and the gas quickly leaves the solution. It comes out in an 'explosion' of bubbles.

ACTIVITIES

11. When you put too much coffee into a cup of hot water, why do coffee grounds appear at the bottom of the cup?
12. You want to mix powdered milk with water. What should you use to make it dissolve more quickly, cold or hot water? Why?
13. If you want to keep the bubbles in a fizzy drink for a longer time, how should you serve it, cold or lukewarm?
14. When you put sugar in a cup of coffee, why do you stir it with a spoon?
15. Write an experimental procedure to prepare a supersaturated solution of salt in water. Compare with a partner and try the procedure in the lab.

IMPORTANT

Saturation of a solution can happen: with a very little amount of solute, when the solute is insoluble, or with a large amount, when the solute is very soluble.

IMPORTANT

Supersaturated solutions are unstable. A slight stir can cause the precipitation of all the excess dissolved solute.

DID YOU KNOW?

Many fizzy drinks have carbon dioxide (CO_2) gas dissolved in them. This comes out in the form of bubbles when you open the bottle.

This video shows the industrial manufacturing of carbonated drinks:

goo.gl/43cWFC

4 Concentration of solutions

SCIENCE 2.0

This video illustrates the concentration of solutions:

goo.gl/qGFwFE

As we have studied, solutions are mixtures of one or several solutes and a solvent. But from this perspective, we are only qualitatively analysing them. We must learn a little more in order to work with them in the laboratory or in industry.

THINK AND REASON

You know that there are different types of solutions depending on the relative proportions of the components. But as a science student, you need to know a little more about solutions.

What more is needed? How do you think you can express this more precisely? Is there only one or are there several ways to express this?

To know about the specific composition of a solution we will use the concept of **concentration**. This quantity expresses the ratio between the amount of solvent and the amount of solute or solution.

WORKSHEET

Download this unit's worksheet and do activities 12 and 13.

4.1. Mass percent concentration

The **mass percent** is expressed by relating the mass in grams of dissolved solute for every 100 grams of solution.

$$\text{Mass percent concentration} = \frac{\text{mass (g) of solute}}{\text{mass (g) of solute} + \text{mass (g) of solvent}} \cdot 100$$

or also expressed like this:

$$\text{Mass percent concentration} = \frac{\text{mass (g) of solute}}{\text{mass (g) of solution}} \cdot 100$$

SOLVED EXAMPLE

1. **Calculate the mass percent concentration of a solution of 20 g of sodium chloride (NaCl) in 0.3 L of water (*d* = 1 kg/L).**

Solution

First calculate the mass of the solvent:

$d = \frac{m}{V}$. Solve for the unknown 'm' and we are left with:

$$m = d \cdot V = 1\ \frac{\text{kg}}{\text{L}} \cdot 0.3\ \text{L} = 0.3\ \text{kg} = 300\ \text{g}$$

$$\text{mass}_{\text{solution}} = m_{\text{solute}} + m_{\text{solvent}} = 20\ \text{g} + 300\ \text{g} = 320\ \text{g}$$

Therefore: $\text{Concentration} = \frac{20\ \text{g}}{320\ \text{g}} \cdot 100 = 6.3\ \%$

4.2. Volume percent concentration

When the solutions are liquid the easiest thing to do is to express the concentration in **volume percent.**

THINK AND REASON

Remember what you have just learnt in the previous section. What expression should we use if we want to express the concentration of the solution in volume percent?

SOLVED EXAMPLE

2. Calculate the concentration expressed in volume percent of a solution prepared with 200 mL of ethyl alcohol, which has had water added to complete 500 mL of solution.

Solution

We have 200 mL of ethyl alcohol in 500 mL of solution:

$$\text{Volume percent concentration} = \frac{\text{Volume of solute}}{\text{Volume of solution}} \cdot 100$$

$$= \frac{\text{200 mL of alcohol}}{\text{500 mL of solution}} \cdot 100 = 40\,\%$$

That is, there is 40 mL of alcohol in every 100 mL of solution.

4.3. Concentration in grams per litre

Another common way of expressing concentration is to consider the mass of the solute expressed in grams per each litre of solution.

The mathematical expression used is:

$$\text{Concentration (g/L)} = \frac{\text{Mass (g) solute}}{\text{Volume (L) of solution}}$$

ACTIVITIES

16. How many grams of mass of a 10 % sodium chloride (NaCl) solution are necessary to have 20 grams of pure NaCl?

17. The solubility of potassium iodide (KI) at 60 ºC is 200 g/L. What mass of potassium iodide do we have to use to prepare 350 mL of saturated solution at that temperature?

18. Calculate the concentration, expressed in g/L, of a solution prepared by mixing 25 g of table salt (NaCl) with 500 g of water. If the final volume of the solution is 511 mL, what is its density?

19. If a spoonful of sugar is 3 g and a glass of water has a capacity of 200 mL, how many spoonfuls must we add to have a solution concentration of 75 g/L?

5 Techniques for separating mixtures

Separating heterogeneous mixtures

Filtration

An insoluble **solid** in a **liquid** medium.

First prepare a mixture of water and sand in a receptacle.
Next, take a bit of filter paper, a funnel and a glass. Gradually pour the mixture through the funnel where you have placed the filter paper.

Describe what you observe. Is the liquid you obtain in the glass a solution? The sand collected in the filter paper is wet. Find a partner and think about how you could have the sand back as it was before preparing the mixture.

Sc

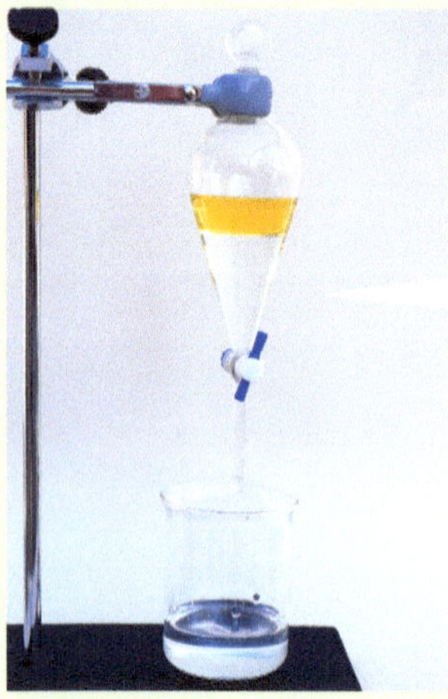

Decantation

Two or more **liquids** that do not mix together.

Prepare a mixture of water with oil in a receptacle.
Next, pour it carefully into a separating funnel. When the separation between these liquids is clear, open the valve and the denser liquid will begin to fall, in this case, the water.
When almost all the water has fallen, we change the glass. We throw out the first portion and begin collecting the oil.

Describe what you observe. What is the density of each substance?
Why does the denser substance stay below the other substance?
Discuss your observations with a partner.

Liquid

Oth

Centrifugation

Place a mixture of water and sand in a device called a centrifuge. With a rotational movement, it makes the particles with a higher density end up at the bottom with the liquid at the top.

Describe what you observe. Why does the sand end up at the bottom and the water above?

Magnetic separation

Other situations

Mix sulphur and iron filings until you have a uniform mixture.
Cover the magnet with paper and pass it over the mixture. Change the receptacle and take the magnet out carefully.

Describe what you observe. What characteristic must be present for this technique to work?

Separating homogeneous mixtures

Crystallisation

A **soluble** solid in a **liquid** medium.

In these situations, the compound that we wish to obtain must be soluble in a hot solvent and not very soluble in the same solvent when cold. The impurities that we want to separate are insoluble in this solvent.

Take 1 g of copper sulphate and pour it into a receptacle with water. Heat this so that it completely dissolves. Filter the impurities that have not dissolved.
While it is still hot, put the solution in a crystalliser and let it cool. As the temperature decreases, the solution will become supersaturated and crystals of the substance we want to purify will appear.

Describe the experiment. Why does the solution become saturated as it cools?
Discuss the relationship with solubility of copper sulphate in water with a partner.

id

Distillation

Two or more **liquids** that mix together.

This is based on different boiling points of the substances.

Liquid

Make an assembly like the one in the figure. Put a mixture of water and alcohol in the flask. As the temperature rises, observe how you obtain alcohol in the distilling flask.

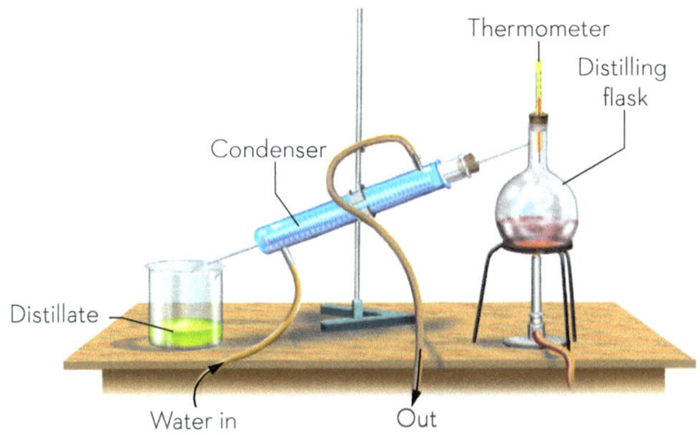

Describe what happens in the experiment. What is the temperature at which the mixture starts boiling? What is the condenser for?

tuations

Other situations

Chromatography

This technique is used to identify the components of a solution, not for separating them. It is based on the different diffusion rates that different substances have in the same solvent.

Take a 15 x 10 cm sheet of filter paper and draw a line 2 cm from the larger side. Draw circles on this with pens and mark the colour used with a pencil below it. Close the paper by forming a tall cylinder and join it with a staple or a clip. Prepare a 10 % alcohol solution in water and put the cylinder inside without touching the sides (very important). Leave it inside for 15 minutes, then take it out and let it set.

Describe what you observe. Why do the different substances of the ink go up the paper? Could you do the same experiment with water?

MIND MAP

Download and print the worksheet you will find in the OLC to complete the mind map of the unit.

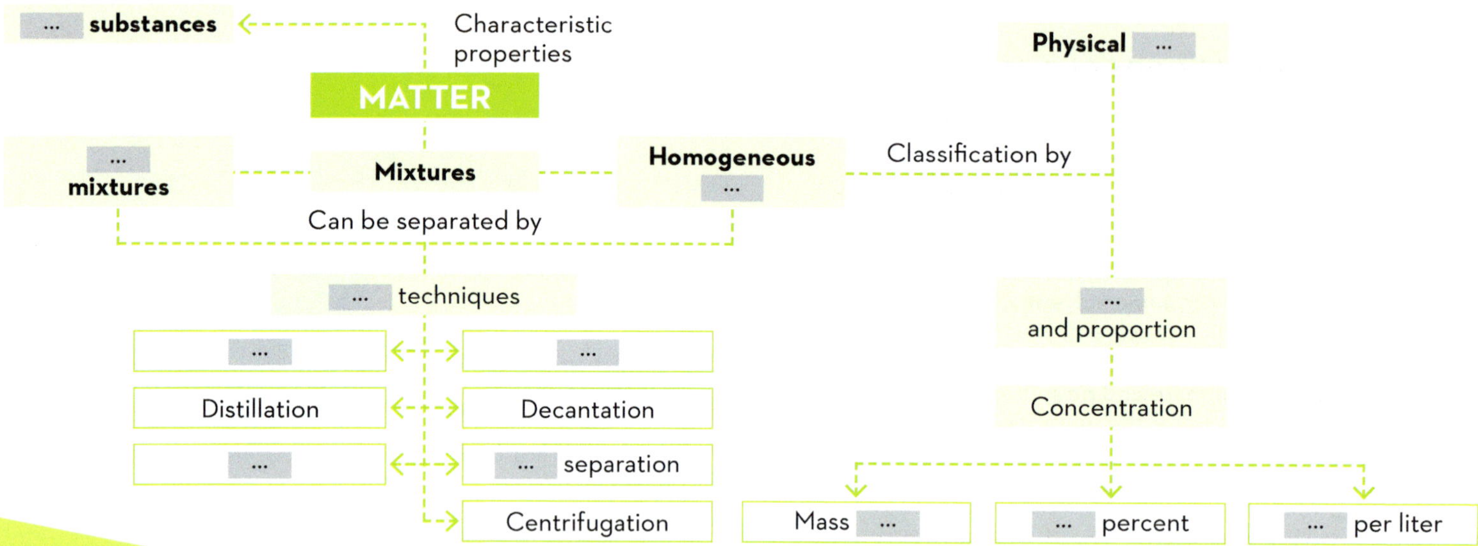

LOOK AROUND YOU

How does alcohol affect driving?

Drinking before driving is one of the main causes of accidents. Alcohol produces alterations in behaviour and reduces reflexes, so it affects almost all the abilities we need for safe driving.

Alcohol and driving: of every 100 accidents, between 30 and 50 have to do with alcohol.

With a **blood alcohol concentration of 0.5 g/L** and higher, the effects on most people are evident. **Above the legal level**, drivers are usually not aware of the risk and **do not take appropriate precautions.**

The **effects** alcohol can have on drivers are varied depending on the blood alcohol concentration.

- People who have drunk alcohol underestimate the effects and alterations that the alcohol produces, typically saying 'I'm in control.'
- Their sense of responsibility decreases.
- Their field of vision decreases and their reaction time increases.

When you consume alcohol, the most dangerous change is the increase in reaction time, decision time (dodge a car or stop abruptly) and response time (execute the manoeuvre chosen).

Irene Mendoza, 07.11.2013. Mapfre-motor magazine.

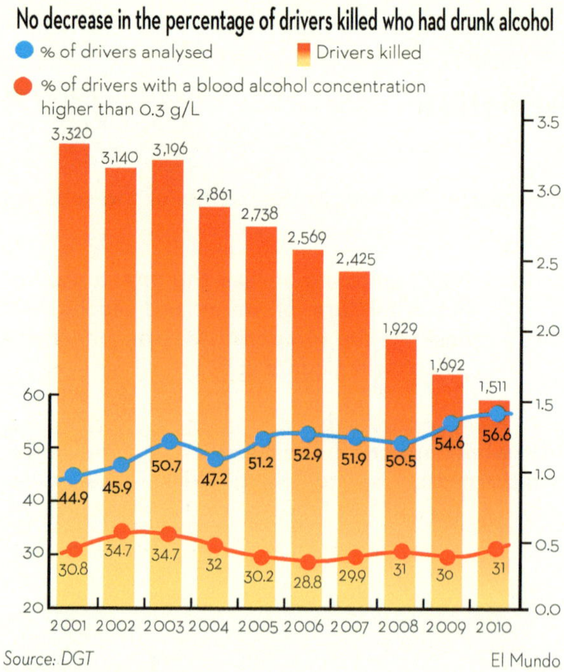

Questions

a) How many drivers killed had a BAC of more than 0.5 g/L between 2005 and 2010? What is expressed by 0.5 g/L?

b) Write an example of the alterations caused by drinking alcohol.

Reading techniques

Use the worksheet from the OLC to practise reading techniques with information extraction strategies.

LABORATORY WORK

Preparing solutions of different concentrations

Objective

Working in small groups, prepare solutions of different concentrations.

Material

- 500 mL volumetric flask.
- 100 mL flask.
- Chemical compounds:
  - Potassium chloride (KCl).
  - Sodium chloride (NaCl).
  - Copper (II) sulphate ($CuSO_4$).
- Scale.
- Beaker.
- Distilled water.

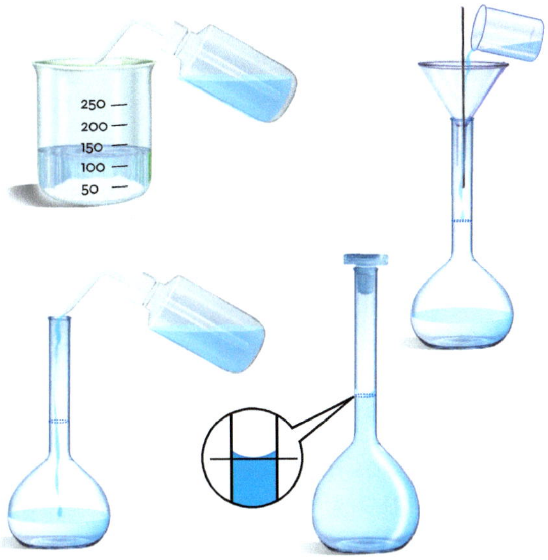

Procedure

We will prepare different solutions:

a) 500 g of 2 % mass percent sodium chloride.

b) 100 g of 3.5 % mass percent potassium chloride.

c) 100 mL of 10 g/L copper (II) sulphate.

Step by step

1. Calculate the number of grams that we need of each of the compounds.
2. Weigh them on the scale.
3. Dissolve them in a little distilled water in the beaker.
4. Pour the solution prepared in the flask and add enough water to almost complete the volume required for each solution.
5. Finally, cover the flask, shake it and add water to fill up to the mark.

Questions

1. Indicate the amounts of each solute that you will use. Copy and complete the following table in your notebook for this.

Solution	Amount of solution (g or mL)	Amount of solute (g)
2 % sodium chloride	...	...
3.5 % potassium chloride	...	...
10 g/L copper sulphate	...	...

2. What does the mark on the flask mean?
3. Express solutions a) and b) in g/L and c) by mass percent.

FINAL ACTIVITIES

Basic activities

1. Indicate which of the following mixtures are homogeneous and which are heterogeneous: a fizzy drink, herbal tea, fruit milkshake, a cup of liquid chocolate and a bowl full of sand and lentils.

2. Give examples of some physical or chemical properties that are different from those proposed in section 1 of the unit.

3. Indicate which of the following mixtures are colloids: blood, saliva, shaving cream, jelly, DNA and ammonia in water. Discuss your answer with a partner.

4. Look for information on the internet about the range of particle size for homogeneous mixtures, heterogeneous mixtures and colloids.

5. In groups of three, brainstorm why some substances are not able to dissolve others.

6. Indicate which type of solution has a higher proportion of solute: in a saturated solution or in a very concentrated solution.

7. Are there solutions with more than one solute? What about with more than one solvent? Work with a partner to find some examples.

8. Calculate the amount of a 15 % by mass potassium chloride (KCl) solution that is needed to obtain 12 grams of pure solute. What if it were a sodium chloride (NaCl) solution?

9. Calculate what percent by mass there is of each component of a solution made up of 350 mL of water and 15 grams of sugar.

10. Can a solution be dilute and saturated at the same time?

11. If the solubility of sugar in water is 200 g/L at 20 ºC, what amount of sugar can we dissolve in 300 mL of water at the same temperature?

12. Write the difference between a saturated solution and a supersaturated solution. How can you transform a supersaturated solution into a saturated solution?

13. If you wanted to separate a mixture of iron powder with water, what method would you use?

14. From the graph, indicate the solubility of potassium chlorate at 80 ºC.

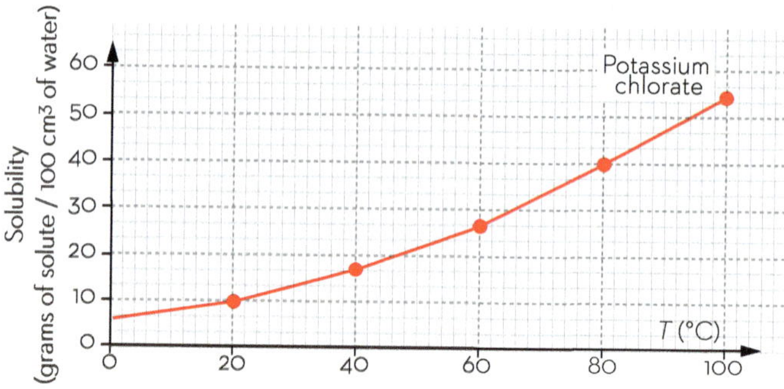

15. In order to treat a wound, the doctor told you to prepare a solution of 35 % alcohol in water. If the alcohol in the first aid kit indicates it is 70 %, how would you do this?

16. Which of the following techniques for separating mixtures allow you to work with all the components once you have separated them?
    a) Chromatography.
    b) Recrystallisation.
    c) Decantation.

17. Look up information on the internet and indicate three different important alloys that were mentioned in the unit. Compare your list with the whole class and make a compilation of contributions.

18. In an experiment you distil a solution of table salt (sodium chloride) in water. What remains in the distilling flask once you have separated all the water?

19. What method of separating mixtures is actually occurring in lands that have salt marshes?

20. We saw in the text that solubility increases with the temperature. Search on the internet for solubility values of sodium chloride (NaCl) at different temperatures. Based on this, should salt be sprinkled on food when it is hot or cold?

Consolidation activities

21. Why does the gas of a carbonated drink come out of the bottle more violently if you shake it first?

22. Manufacturers of clothes irons say that it is better to use distilled water rather than tap water in irons. Why do you think they recommend this?

23. You probably have seen spoilt milk that has gone off. What is the meaning of these degrees? What phases can you observe in it?

24. The amount of alcohol in drinks is expressed in degrees. What is the meaning of these degrees? What volume of pure alcohol will you drink if you have 200 mL of wine (12 % Vol)?

25. Explain in detail how you would prepare 200 mL of a 12 % by mass sodium nitrate ($NaNO_3$) solution if you have a 100 mL flask.

26. You probably have observed that when you take clothes out of the washing machine, they are wet but not dripping. Why do you think this happens?

Advanced activities

27. Throughout the day we do activities like these: *a)* add salt to soup, *b)* add sugar to milk, *c)* make coffee, and *d)* make mayonnaise. Indicate which of these activities involve solutions and which involve colloids.

28. You have a 20 % by mass salt solution in water. If you take 100 grams of it, what amount of salt would you obtain when the solvent is evaporated?

29. Describe a procedure to prepare 100 mL of a 2.5 g/L NaOH solution in water assuming you only have a scale with a 0.1 g sensitivity.

30. If you have a 30 % solution by mass, what would be the new concentration if you add enough water to double its volume?

31. Household ammonia is a solution of ammonium gas in water. Calculate the amount of pure ammonium contained in a bottle of household ammonia that has a volume of 0.25 L if its concentration is 3 % by volume.

32. You have a 25 % sodium carbonate (Na_2CO_3) solution in water. Knowing that the density of the solution is 1.10 g/mL, what concentration is its equivalent in g/L?

33. What amount of alcohol should you add to water to obtain 350 mL of 15 % by volume alcohol in water solution?

34. Work in small groups. Look up information on the internet and make a poster with the different forms of environmental pollution caused by colloids.

35. Sea water is a solution of table salt in water. In some places, like islands, they must desalinate sea water to obtain drinking water. Search for information and explain the techniques commonly used to do this.

Purification of drinking water

Water from rivers is exposed to pollution by materials and microorganisms and requires complex treatment. The turbidity, mineral content and degree of pollution vary depending on the period of the year. (For example, in summer river water is more turbid than in winter.)

The process of purifying drinking water has to go through various stages before the water can be used in our homes.

The stages of this process are: **collection, piping, addition of chemical products, flocculation, sedimentation, filtration and disinfection.**

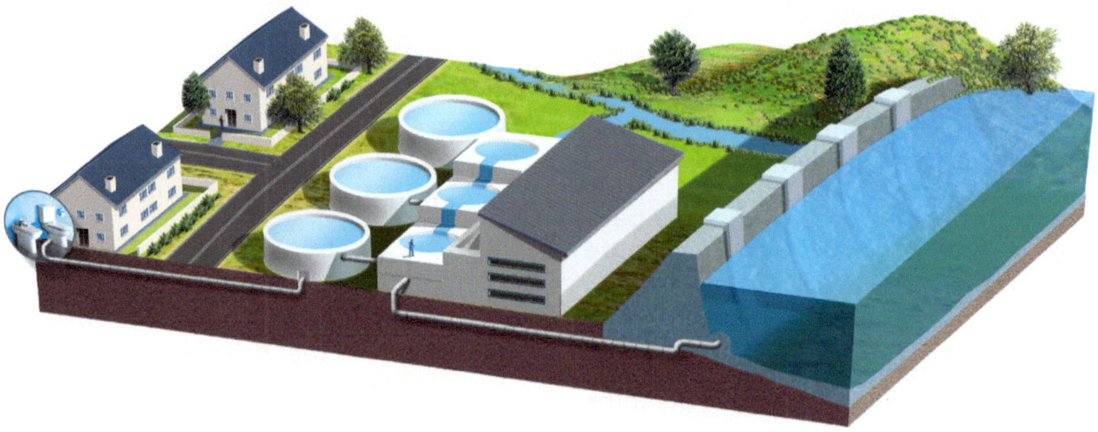

Working in groups, answer the following questions:

Question 1

The collection of surface water is carried out through water intakes that are built in rivers or dykes, but how do you think groundwater is collected?

Question 2

From the intake, the water is piped through aqueducts or open channels. This stage is carried out in pools designed to retain sedimentable solids (sand), the heavy solids that fall to the bottom. Later, the addition of chemical products (coagulants or flocculants) is carried out for the destabilisation of the colloid or turbidity of the water. This process is made by flocculants. Can you explain how coagulants work in this stage of the process?

Question 3

Next, sedimentation is carried out in decanters. The flocculate precipitates to the bottom, forming mud and then it is decanted. Filtration occurs by pouring the sedimented or decanted water over a filter. Once it is filtered, the water is disinfected with different methods. The most commonly used method is by adding liquid chlorine.

Explain how chlorine works to disinfect the filtered water. Do you know other disinfectants commonly used for this process?

Question 4

Working in fours, create a poster showing a diagram of all the stages of the process of purifying drinking water that we have just described. Use a flow chart layout, with a box for each process and connecting lines. You can find one at: **goo.gl/gjcBr**.

Environmental pollution: a problem of mixtures

→| Competency task

Solid, liquid and gas pollutants

The **environmental problems** that we have in the world around us are caused by solid, liquid or gas particles that mix together to form heterogeneous systems or even solutions. These change the properties of the substances that are crucial for survival.

Solid pollutants

When pollutants mix with water and soil they damage flora and fauna. They are insoluble liquids or solids of natural origin and different synthetic products that are thrown into the water as a result of human activities. These include batteries, paint, grease, tar, paper, plastics, glass, textiles, tins, detergents, pesticides, heavy **metals** (like **lead**, **copper**, **zinc**, **tin**, **chrome** and **aluminium**), sulphur and nitrogen by-products, etc.

Liquid pollutants

The substances that mix with water are mostly waste products from industry, agriculture and fishing, commercial establishments, etc. These can eventually lead to certain volumes of polluted groundwater, surface water and/or rainwater. They include organic waste, dead and decomposing animals, petroleum and its by-products, sulphuric acid, oils (engine, used household, etc.).

Gas pollutants

Air pollutants are generated mainly by **industrial combustion** processes, residential heating, urban transport, rubbish burning, forest fires, and others. Among the main effects of this pollution is acid rain, the greenhouse effect and damage to the ozone layer that surrounds the Earth. The main pollutants that dissolve in air are **carbon dioxide** and **carbon monoxide**, **nitrogen oxide** and **sulphur oxide, CFCs, hydrocarbons, ozone** and **ashes.**

Objective

Obtain information on the environmental pollution around us.

Final product

Make a video or a presentation about this issue. Divide the task among the members of the group so that each one can inform the rest of the class about one type of problem detected and its solution.

Step by step

1. Look up the main pollutants that are around you.
2. Indicate how each one of them affects you in regard to health.
3. Obtain information about the methods of environmental management that your city uses.
4. Research the most polluted cities in Spain. Differentiate the type of pollutant that most affects them.
5. Prepare the presentation for the rest of the group with all this data.

Self-assessment

Evaluate your work and your classmates' work with the guidelines that you can download from the OLC.

4 THE STRUCTURE OF MATTER

You have probably seen how a building is constructed. It has foundations, beams, floors, windows, walls, pipes, electrical wiring, etc. If we observe any type of matter, we can see that something similar occurs: it has an internal structure to build it piece by piece.

The smallest particles form atoms. Atoms form simple molecules or more complex structures. These structures often group together in even more complex frameworks to form everything around us. Everything, including humans, is matter.

> 'All things are made of atoms'.
>
> Richard Phillips Feynman

WORK WITH THE TEXT

We can indicate the composition of an object with the phrasal verb 'is made up of'. Look at the example:

A football team is made up of eleven players.

Find at least four examples in the text where you can use this grammar structure and rewrite them in the same way.

CHALLENGE

You have noticed that different substances have different properties. These differences also show up in the way that two substances interact with another one.

Take an ice cube and pour some sand on it. What happens to the sand? And to the ice? Do they mix? Record your observations.

Now pour some table salt on another ice cube. What happens now? Is it different from the sand experiment? Do they mix? Again, record your observations.

This different behaviour is caused by the different chemical bonds in sand, table salt and ice. You will learn about these bonds in this unit.

Contents

KEYWORDS

Download and print the activities that will guide you through the essential vocabulary of this unit.

FIND A CLASSMATE

Download and print the worksheet 'Find a classmate: symbols and names'. Share experiences about this topic with your classmates and discover interesting things about them.

1 Pure substances: simple and compound substances

WORKSHEET

Download this unit's worksheet and do activities 1 and 2.

In the previous unit we learned how to distinguish between pure and mixed substances. But what are pure substances? Do we use them in our everyday lives or only in the laboratory? We are going to identify different types of pure substances. We will see what they are and how they appear.

THINK AND REASON

Look at the following substances:

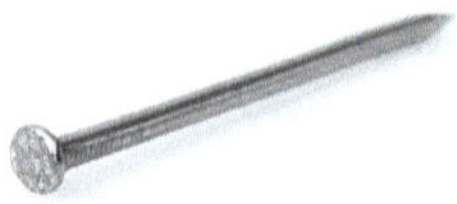
Iron nail

Copper wire

Water

Aluminium saucepan

Air

- Are they pure or mixed substances?
- Which ones are simple substances and which ones are compounds?
- How can you find out?

CHEMISTRY 2.0

Look at the link below to learn some facts about mixtures, pure substances, elements and compounds:

goo.gl/65CuQQ

As you learnt in the previous unit, pure substances have physical properties and chemical characteristics that allow us to differentiate them from other substances.

But there are two possibilities in pure substances:

Simple Substances	Compounds
Simple substances cannot be separated into other substances because they are made of just one element. For example: lead, calcium, zinc, gold, etc. 	Compounds are pure substances which can be separated into other simpler substances. This is because they are made of different elements. For example, ammonia, sulphuric acid, salt, pyrite, etc. 

ACTIVITIES

1. **Indicate whether the following are simple substances or compounds: sodium chloride, sugar, copper wire, oxygen gas, and water. Compare your answers with a partner.**

2 Atoms and subatomic particles

In ancient Greece, the philosophers Leucippus and Democritus believed that matter was made of small particles and therefore could not be divided indefinitely. They thought there had to be an inseparable or indestructible block. They called these particles **atoms,** which means 'indivisible'.

This idea was put aside by scientists until 1808, when **John Dalton** formulated the hypothesis that matter was composed of elemental units that could not be altered or divided. He also called them atoms.

Nowadays, with the help of special microscopes, we can see individual atoms and even interact with them.

THINK AND REASON

Imagine you have a coffee bean.

- Can you say approximately how big it is?
- What about a grain of rice, how long is it?
- Can we continue looking for smaller entities indefinitely?
- How far do you think you could get?
- What is the smallest object you can think of?

The size of matter can be measured. Here are some examples of increasingly smaller objects:

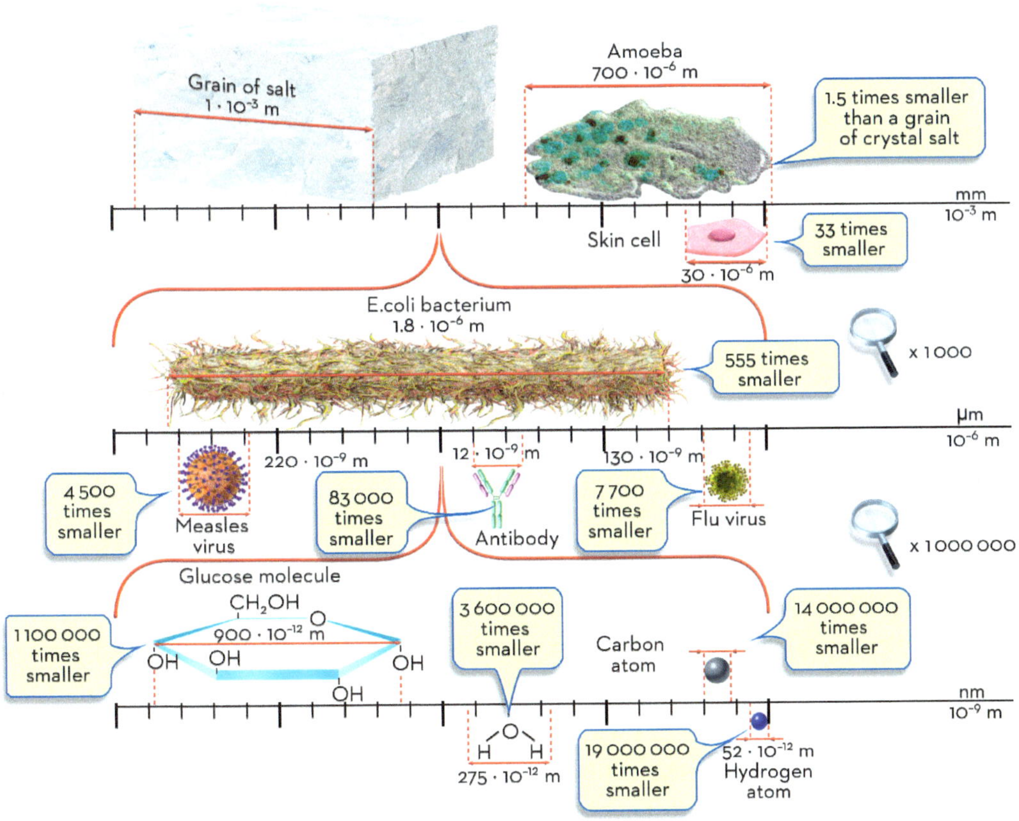

SCIENCE 2.0

You can watch a video on the evolution of the size of matter, from the largest to the smallest, at:

goo.gl/oxHHL

SCIENCE 2.0

This animation takes you on an incredible journey from the macroscopic world to the microscopic:

goo.gl/NwjaVC

WORKSHEET

Download this unit's worksheet and do activity 3.

2.1. Atoms

We now know that matter can be split into smaller and smaller parts until we reach atoms, which are invisible to the naked eye.

> An **atom** is the smallest particle a simple substance can be divided into without losing its chemical properties.

A chemical element is a substance that cannot be separated into simpler substances by any physical or chemical process. In other words:

> A **chemical element** is a substance made of only one type of atoms.

For example, carbon only has carbon atoms and oxygen only has oxygen atoms.

As we have seen, matter is neutral. Therefore, common sense tells us that atoms are also neutral.

EXPERIMENT

1. You will need: a balloon inflated with air, a fizzy drink can and a piece of woollen fabric. Rub the balloon with the woollen fabric and put it near the can without touching it. Describe what occurs.

- Why do you think the can moves towards to the balloon?
- What type of particles must the can and the balloon have at this instant?
- What happens if you put the balloon close to the can without rubbing it with wool?
- If the can and the balloon are made of neutral particles, what consequence comes from our experiment?

This experiment confirms that atoms are made up of particles that are electrically charged. They are **subatomic particles.**

IMPORTANT

Symbols for some elements:

Hydrogen. . . . . H
Carbon C
Nitrogen. . . . . N
Oxygen. . . . . O
Fluorine . . . . F
Sodium . . . . Na
Chlorine Cl
Iron . . . . Fe

WORKSHEET

Download this unit's worksheet and do activity 4.

2.2. Subatomic particles

There are three types of subatomic particles in atoms: **electrons, protons** and **neutrons.**

The electron was the first particle to be discovered. It was discovered by J.J. Thomson in 1897. It is the smallest known negative charge and its value is $-1.6 \cdot 10^{-19}$ C.

Electron (e^-) 

- Negative charge: −1.
- Very small mass ($9.11 \cdot 10^{-31}$ kg).

In 1919 Ernest Rutherford discovered the proton. Its charge is the same as the electron, but positive. It is the smallest positive charge and its value is $+1.6 \cdot 10^{-19}$ C.

Proton (p^+) 

- Positive charge: +1.
- Mass 1 836 times greater than the electron ($1.67 \cdot 10^{-27}$ kg).

Because matter is neutral, atoms must have the same number of electrons and protons. They are atoms with no net charge. In other words, they are neutral.

In 1932 James Chadwick discovered that the atom contains another type of particles he called neutrons. They have no charge and their mass is almost the same as the proton.

Neutron (n^o)

- No charge
- Mass equal to the proton ($1.67 \cdot 10^{-27}$ kg).

So, atoms are not the smallest part of matter. Are the particles they are made of the smallest? We now know that protons and neutrons can be split into smaller particles called ***quarks.***

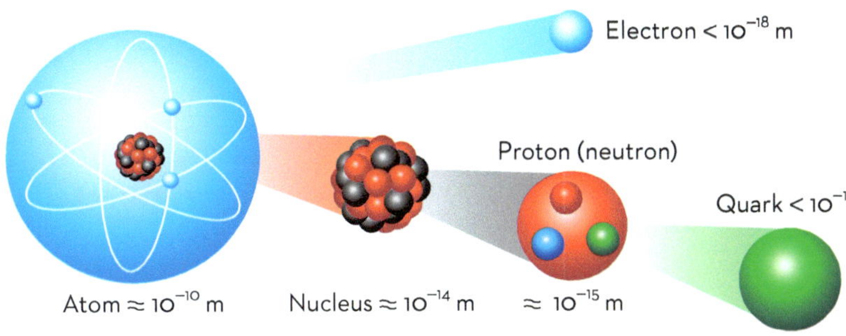

> **IMPORTANT !**
>
> Any **negative charge** is a whole multiple of the **electron's** charge.
>
> Any **positive charge** is a whole multiple of the **proton's** charge.

> **WORKSHEET**
>
> Download this unit's worksheet and do activities 5 and 6.

ACTIVITIES

2. An atom has 3 protons, 3 electrons and 4 neutrons.

a) What is its mass?

b) What mass would it have if it did not have electrons?

c) What charge does the atom have?

d) What charge would it have if it only had two electrons?

3. Indicate which of the following contain atoms:

a) A lump of iron. *b)* A diamond.
c) A ray of light. *d)* Thunder.
e) A bacterium. *f)* Air.
g) Water.

Discuss your answers with a partner. How did you come to your conclusions?

3 Atomic models

You have learnt that atoms are made of protons, electrons and neutrons. In this section, you will learn about their placement inside the atoms.

WORKSHEET

Download this unit's worksheet and do activity 7.

THINK AND REASON

Imagine you have some marbles: negatively charged red ones; positively charged green ones; and white ones with no charge.

- If you took a group of five green marbles, where would you put the white ones so they do not repel each other?
- What charge would the green and white marbles have together?
- If you want to counteract the charge of this group, how many red marbles would you need?

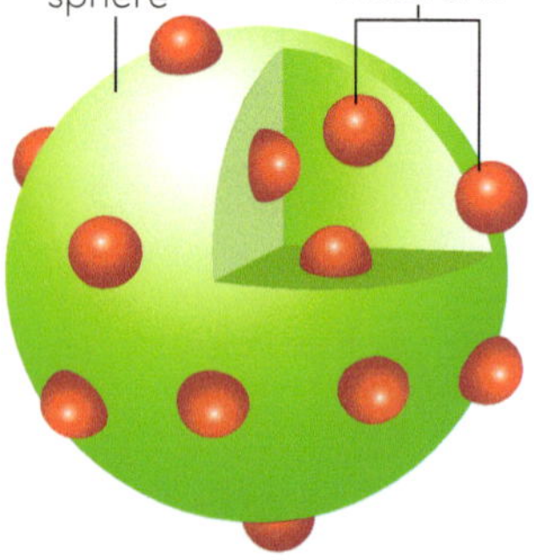

The first **atomic model** was devised in 1898 by **J.J. Thomson.** The model suggested the existence of a **sphere of positive electricity.** They had not yet discovered protons as individual particles. The theoretical sphere included as many electrons as were needed to neutralise it.

This idea was colloquially known as the 'plum pudding model'. Thomson imagined a proton mass in the shape of a sphere with negative granules in it, the electrons. These neutralised the mass and formed atoms. Subsequent experiments proved this model did not match reality.

Successive experiments led to the next model, created by **Ernest Rutherford.** The planets are attracted to the Sun, but do not fall on it. Instead, they orbit around it. Rutherford thought the atom behaved in the same way.

Like planets, electrons orbit around protons and neutrons. The protons and neutrons are in an unchanging cluster, similar to the Sun.

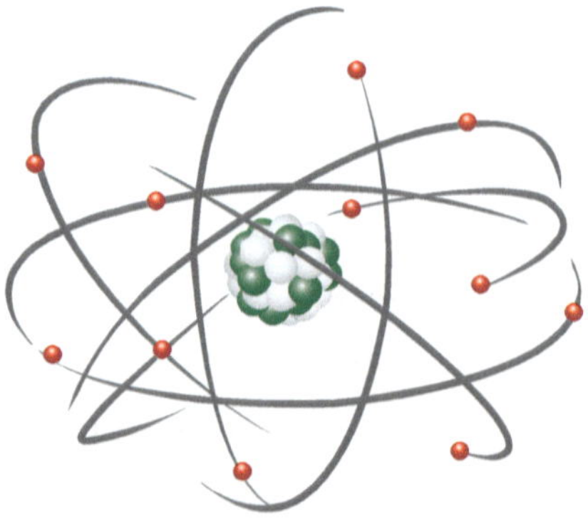

In Rutherford's model, this central cluster is the **nucleus** of the atom. Because the protons have a positive charge, they repel each other. But something must prevent this from happening. The **neutrons** do this: they are in the nucleus as a barrier between the protons, grouping together to prevent them from repelling each other. This means the nucleus is stable.

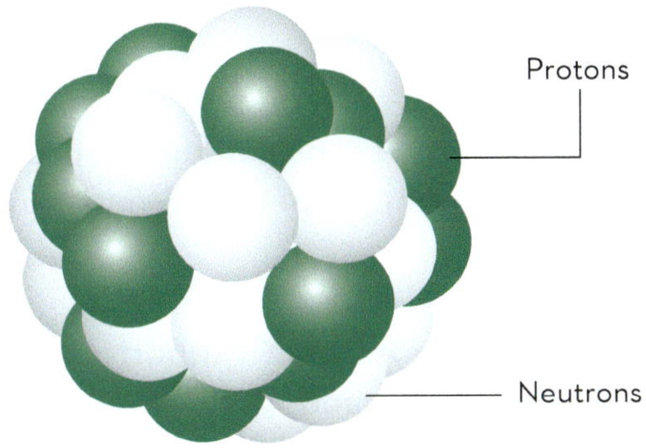

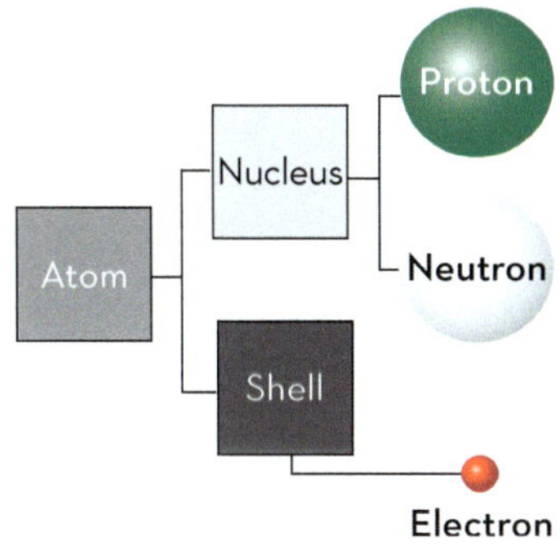

The electrons form the **shell** of the atom, moving around the nucleus. If they were motionless, they would fall on the nucleus due to electrical attraction. As they move, they counteract the force of attraction.

Finally, the Danish physicist **Niels Bohr** discovered that electrons do not move randomly. They revolve around the nucleus in stable orbits. Each orbit has a different number of electrons. The further away an orbit is from the nucleus, the more electrons it can hold, because it will be bigger. Orbits are larger when they are further from the nucleus.

WORKSHEET

Download this unit's worksheet and do activities 8 and 9.

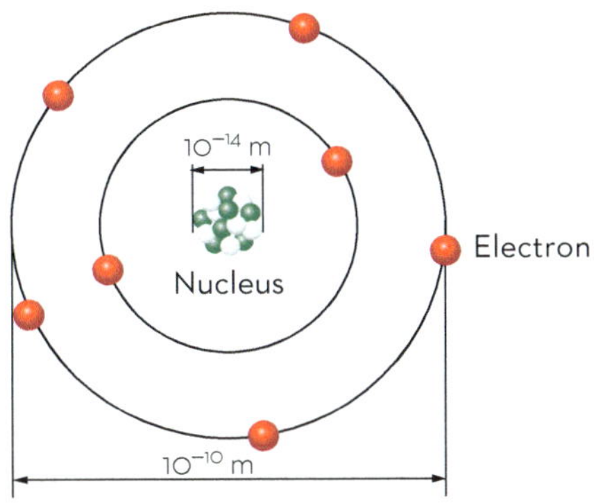

The nucleus measures approximately 10^{-14} m and the electrons move around it at an average distance of 10^{-10} m. This means that they occupy a sphere with a radius approximately 10 000 times bigger than the nuclear sphere. The volume of an atom is a trillion (a million million) times bigger than that of the nucleus.

DID YOU KNOW?

If the nucleus were the size of a football, the shell would be the size of the stadium. The electron would be the size of an olive stone moving around the stadium.

ACTIVITIES

4. Draw an atom with 5 protons and 6 neutrons in its nucleus. How many electrons do you need to add to make it neutral?

5. Do you think an atom can have fewer neutrons than protons? Discuss with a partner.

6. The text above gives the relative sizes of an atom and its nucleus. If the nucleus were the size of a 30 cm ball, how big would the atom be?

4 Atomic and mass numbers

As we have explained, each type of atom represents a certain chemical element. This means that there must be something in the atom that differentiates it from all other atoms.

THINK AND REASON

The element copper has 29 protons, nickel has 28 protons and zinc has 30 protons. Each element has different properties. What conclusion can you draw from this information?

The scientists Henry Moseley and Ernest Rutherford performed several experiments with atoms between 1913 and 1919. These experiments led them to the conclusion that the number of protons atoms have in their nuclei is what characterises them and differentiates them from others.

This crucial quantity of protons in the atom is called the atomic number. It is like the fingerprint for each element: it makes each element unique and identifiable.

> The **atomic number** is the same number of protons there are in the nucleus of an atom. It identifies each element. The symbol for the number of protons is the letter **Z.**

The atomic number is represented by a number in subscript to the left of the element symbol. For example $_{8}O$, $_{17}Cl$, $_{26}Fe$.

When the atom is electrically neutral, the number of electrons is equal to the number of protons. For example, an atom with 6 protons is always the element carbon and it will have 6 electrons if it is neutral.

Because the electrons have a very small mass, the mass of the atom is almost exclusively due to the protons and neutrons it has. This allows us to define:

> The **mass number** is the total number of protons and neutrons there are in the nucleus. The symbol for the mass number is the letter **A.**

The mass number is represented by a number in superscript to the left of the element symbol. For example, ^{16}O, ^{35}Cl, ^{56}Fe.

> Number of protons (Z) + Number of neutrons = Mass number (A)

From this we can deduce that the number of neutrons can be calculated by subtracting the atomic number from the mass number: Number of neutrons = A – Z

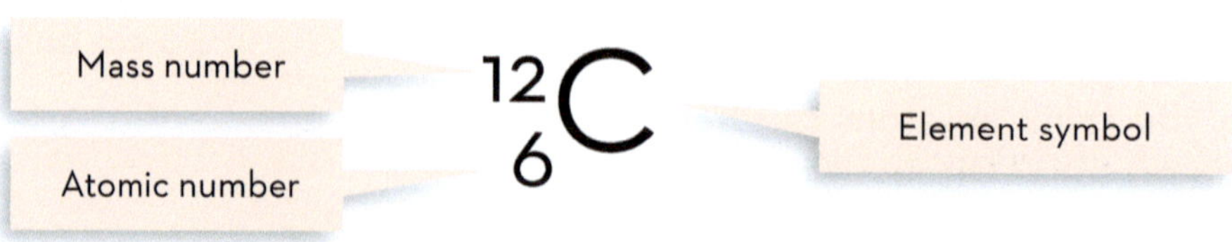

DID YOU KNOW?

To calculate the total mass of the atom we need to add the separate masses of the protons, neutrons and electrons. However, the mass of the electrons is around 1000 times less than the sum of the other two. Therefore, its value hardly affects the total mass.

WORKSHEET

Download this unit's worksheet and do activity 10.

4.1. Isotopes

Some of the particles in atoms can alter their number without the type of element changing. They are the electrons and neutrons, as we will see below.

THINK AND REASON

Atoms are composed of protons and neutrons in the nucleus and electrons in the shell.

- What do you think would happen if we remove protons from or add protons to the nucleus? Would it still be the same element?
- What about if we alter the number of neutrons? Would the type of element change?

Isotopes are varieties of **atoms** with the same atomic number but a different mass number. Therefore, they are the same element, but they have a different number of neutrons in their nucleus.

For example, there are carbon atoms ($Z = 6$) with the following mass numbers: 12, 13 and 14. Therefore, they each have 6, 7 and 8 neutrons respectively. They are isotopes of the same element.

The elements we find in nature are a mixture of isotopes. Most chemical elements have more than one isotope. Only 21 elements have a single natural isotope, such as fluorine, sodium, phosphorus, cobalt and gold. By contrast, **tin** has the highest number of stable isotopes of any element; it has 10.

We identify isotopes by the name (or symbol) of the chemical element followed by its mass number. For example, iron-57, uranium-238 and helium-3.

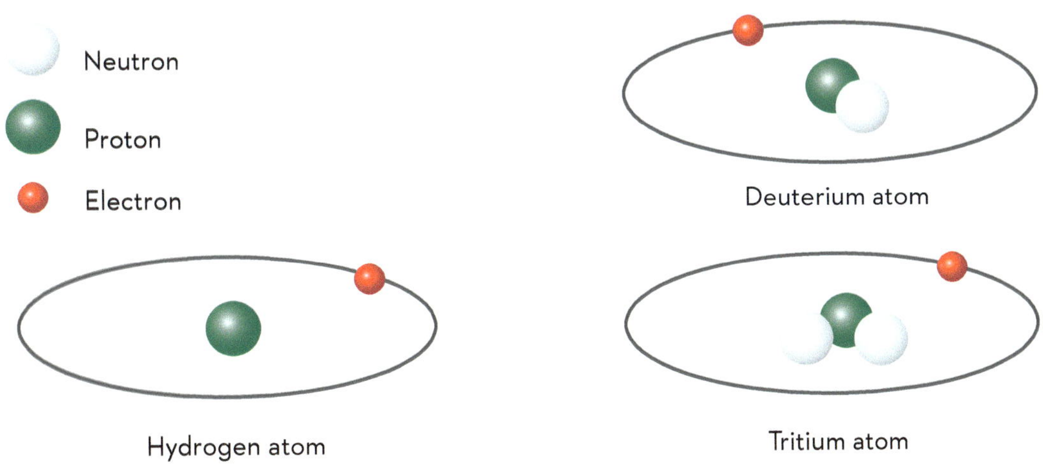

The element hydrogen has three isotopes with a specific name: protium (with no neutrons), deuterium (with 1 neutron) and tritium (with 2 neutrons).

CHEMISTRY 2.0

Here is a link to an interactive exercise to construct atoms and isotopes:

goo.gl/ajxctT

WORKSHEET

Download this unit's worksheet and do activities 11 and 12.

CHEMISTRY 2.0

The following videos show the behaviour of heavy water:

goo.gl/2duLsZ

and the behaviour of isotopes:

goo.gl/Wzxda5

WORKSHEET

Download this unit's worksheet and do activity 13.

CHEMISTRY 2.0

The video at the following link shows the main medical applications of radioactive isotopes:

goo.gl/F7yVfd

THINK AND REASON

We have seen that the element hydrogen has three isotopes. So imagine a molecule of water (H_2O) made of deuterium atoms, which have one proton and one neutron in their nucleus, instead of protium atoms, with only one proton in their nucleus. These molecules exist in something called heavy water.

How can differentiate between water and heavy water?

1. If you have some ice (made from water) and you put it in a glass containing liquid water, what do you think will happen? Suggest an explanation for what happens.
2. If you had ice made from heavy water, what do you think would happen if you put it in a glass containing liquid water? Suggest an explanation.

H_2O
Molecule of ordinary water

D_2O
Molecule of heavy water

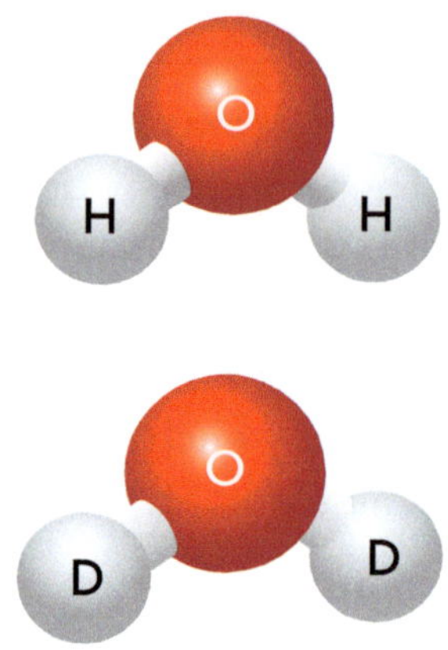

So far, we have described natural isotopes. There are also **artificial isotopes** that are prepared in the laboratory. The number of neutrons in the atoms increases when their nuclei are bombarded with particles containing neutrons.

Many isotopes are unstable and disintegrate quickly. This means they emit particles, breaking apart the initial nuclei and forming others. In many cases, this process creates new elements.

This is very useful in medicine, archaeology and agriculture, among other areas. One extremely important example is the isotope **carbon-14**. We use this to find out the age of the organic remains of different organisms.

ACTIVITIES

7. **The element chlorine has Z = 17 and A = 35. Calculate the composition of its nucleus and how many electrons it has.**

8. **An atom has 20 protons and a mass number of 40. How many neutrons and electrons does it contain?**

9. **Do you think an atom with Z = 7 can have an isotope with Z = 8? Discuss your answer with a partner.**

4.2. Ions

You have seen what happens to the atoms of an element when we modify the number of neutrons. But what happens when we remove or add electrons?

THINK AND REASON

You know that an atom is neutral because it has the same number of protons and electrons.

- What do you think happens to an atom in an element when it gains an electron in its shell? Does its net electrical charge change? Is it still the same element?
- What happens to another atom when it loses an electron? What electrical charge would it have in this case?

Ions are atoms that gain or lose electrons in their shells. They are called **cations** when they are positive and **anions** when they are negative.

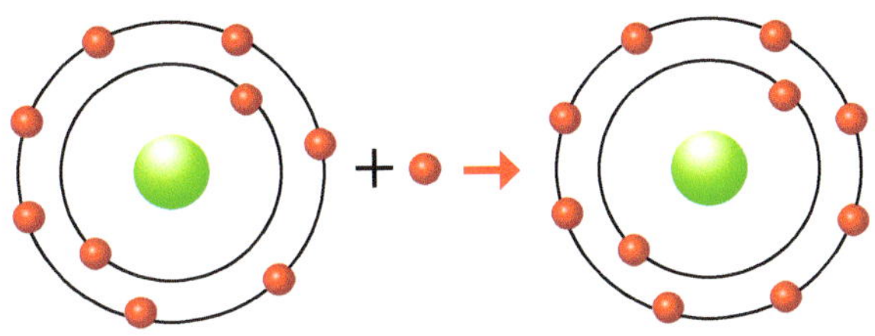

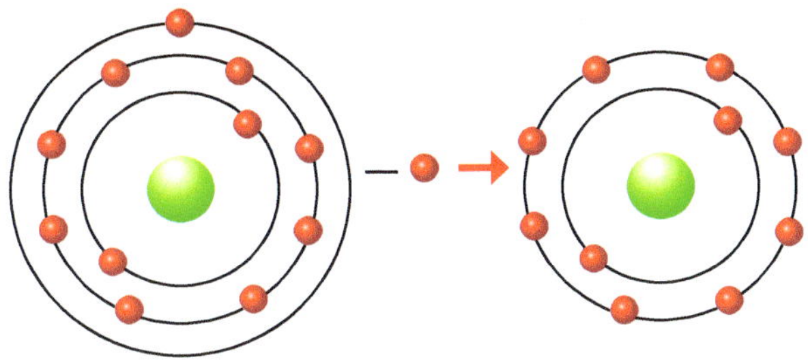

WORKSHEET

Download this unit's worksheet and do activities 14 and 15.

So, when a fluorine atom (Z = 9) gains an electron, it is still fluorine because it still has 9 protons. However, it will have 10 electrons, so its net charge will be –1 (one negative charge more than the number of positive charges). Therefore, we say that we have a fluorine ion (called fluoride) and its symbol is F^-.

A sodium atom (Z = 11) has 11 protons in its nucleus and 11 electrons in its shell. When it loses an electron, it will have a net charge of +1 (one positive charge more than the number of negative charges). We then say that we have a sodium ion and its symbol is Na^+.

CHEMISTRY 2.0

The following video demonstrates the process of ion formation:

goo.gl/NOKkZx

ACTIVITIES

10. How many electrons must the neutral atoms of the elements calcium and nitrogen gain or lose to become the ions Ca^{2+} and N^{3-}?

11. An ion of an aluminium atom (Z = 13 A = 27) contains 10 electrons. Indicate the net charge of the ion and how many neutrons this atom has.

5 Introduction to the periodic system

WORKSHEET

Download this unit's worksheet and do activity 16.

There are many chemical elements, but some of them share properties. This means we can put them in groups. One such property is **electrical conductivity.** This helps us differentiate between metals, which are good conductors of electric current, and non-metals, which are not.

CLASSROOM LABORATORY Metals and non-metals

1. Prepare a circuit like the one below, leaving a space between the two wires. We are going to put different materials made of one chemical element in the space to see if the bulb lights up. You will need: an iron nail, an aluminium fizzy drink can, a lump of coal, a gold ring, a silver spoon, copper cable and a little sulphur powder.

1. Indicate what you observe in each test.
2. Classify the materials into metals and non-metals.

We can group some of their properties into the following table:

	Shine	Physical state (at room *T*)	Conductors of heat	Conductors of electricity
Metals	Yes	Solid (except mercury)	Yes	Yes
Non-metals	No	Solid, liquid and gas	No	No

CHEMISTRY 2.0

Here you will find a video showing the physical and chemical properties of metals and non-metals, as well as their most common uses:

goo.gl/3mA4vi

ACTIVITIES

12. **Find five elements which are solid in their natural state. Indicate whether they are metals or non-metals. Work as a group and find other properties characteristic to metals.**

13. **One of the properties of metals is that they are ductile. (They can be deformed by the action of a force without breaking.) Based on this property, indicate whether the following are metals or not: aluminium, carbon, wood, lead and gold.**

Grouping elements into **metals and non-metals** is very simplistic and does not allow us to order them in a more detailed way. We need to find another method that permits better classification.

Elements are grouped in this way:

According to two criteria

- Increasing atomic number → Arranged in **rows** called **PERIODS**
- Similarity of physical and chemical properties → Arranged in **columns** called **GROUPS**

The **periodic table** has a peculiar shape. As you will see, this is so that elements with similar properties are placed into the same group (or column).

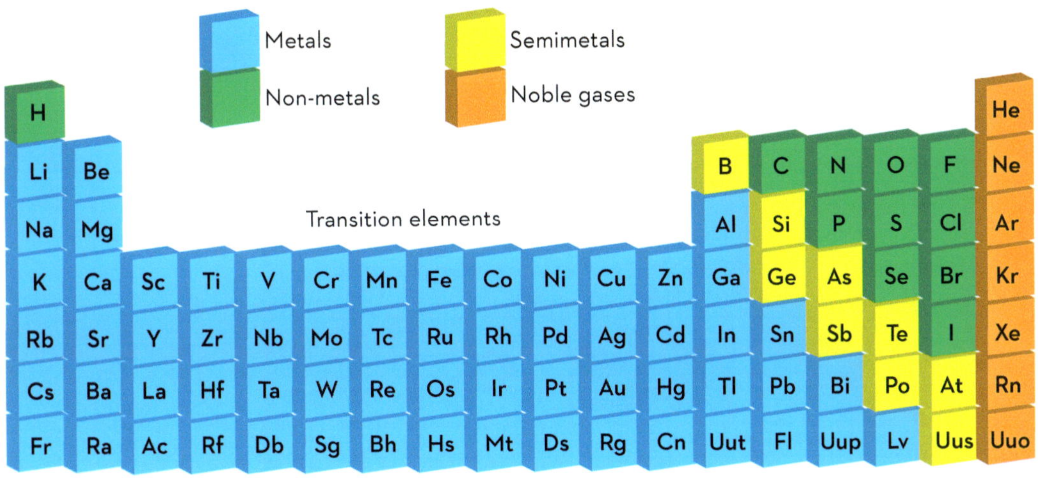

ACTIVITIES

14. Indicate the full name of the following chemical elements: S, K, Ag, Au, Mg, Al, Ca, P, Cu, Zn and Hg.

15. Indicate the symbol and the natural physical state of the following elements: helium, carbon, oxygen, silicon, silver, mercury, cobalt and potassium.

16. Work with a partner to find out which are the four most abundant elements in living organisms.

WORKSHEET

Download this unit's worksheet and do activities 17 and 18.

IMPORTANT

The elements in the column on the far right are inert gases. They have chemical properties that are very different from the rest of the elements in the periodic table.

6 Chemical bonds: molecules and crystals

! IMPORTANT

Hydrogen also combines with non-metals with a covalent bond.

Compound pure substances are made of various types of different atoms with a stable union between them. We call the unions between these atoms **bonds**.

However, not all elements form bonds. The atoms of a few elements called **noble** or **inert gases** do not bond to others. The inert gases are in the far right-hand column of the periodic table. The most important ones are: helium, argon and neon.

Bonds are formed between the same atoms in different quantities or between different atoms.

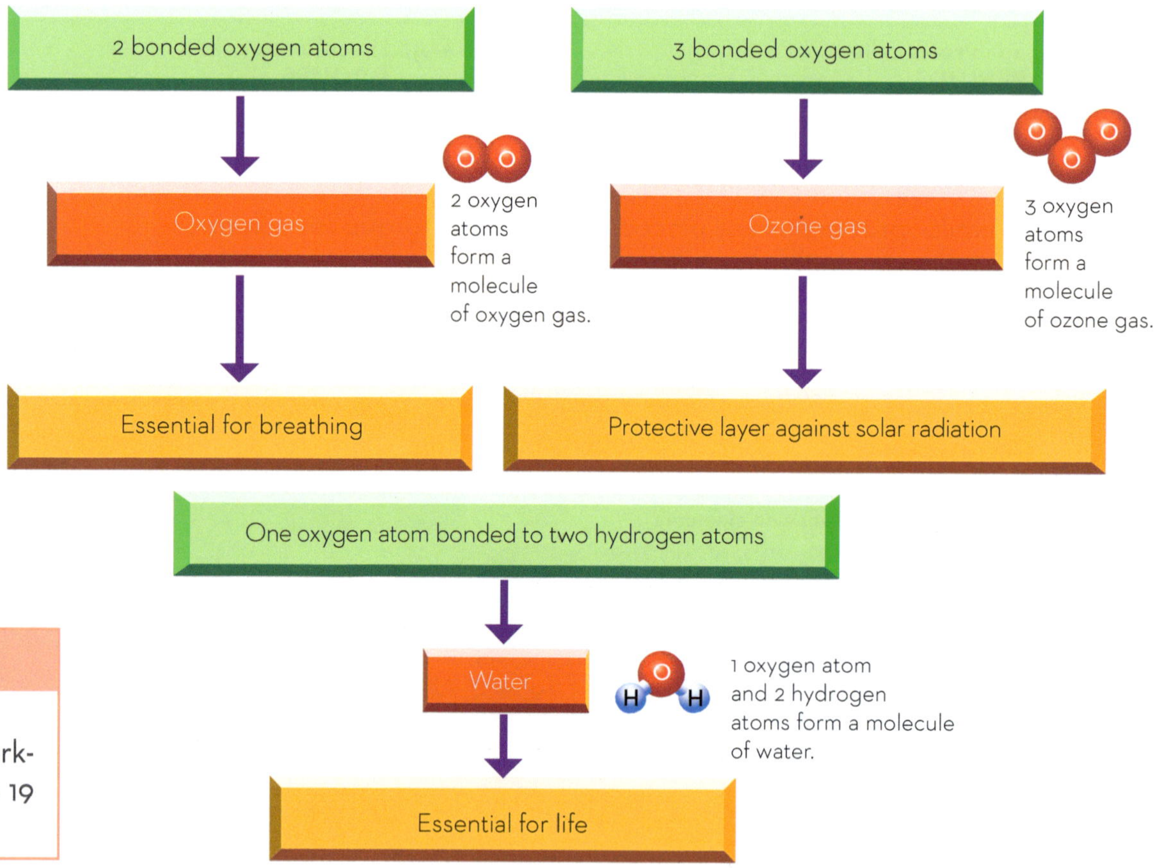

WORKSHEET

Download this unit's worksheet and do activities 19 and 20.

The table below shows the different types of bonds that form depending on the type of atoms which bond. It also shows their resulting chemical form:

Type of bond	Atoms which bond	Resulting structure
Covalent	Non-metal with non-metal	Molecule / crystalline lattice
Ionic	Metal with non-metal	Crystalline lattice
Metallic	Metal with metal	Crystalline lattice

6.1. Molecules

Molecules form when a number of non-metal atoms form a bond.

A **molecule** is the smallest unit of a pure substance that conserves all its properties.

There are small molecules, made of a small number of atoms, and larger ones, which contain a higher number of atoms. The latter are called **macromolecules**. Proteins are one example; they contain hundreds of atoms.

Many molecular substances are liquids or gases at room temperature. Some can be solid, but they melt at low temperatures. Some examples are:

State
- Solid → Examples: iodine, sulphur, phosphorus.
- Liquid → Examples: water, alcohol, acetic acid, acetone, oil.
- Gas → Examples: oxygen, nitrogen, ammonia, butane.

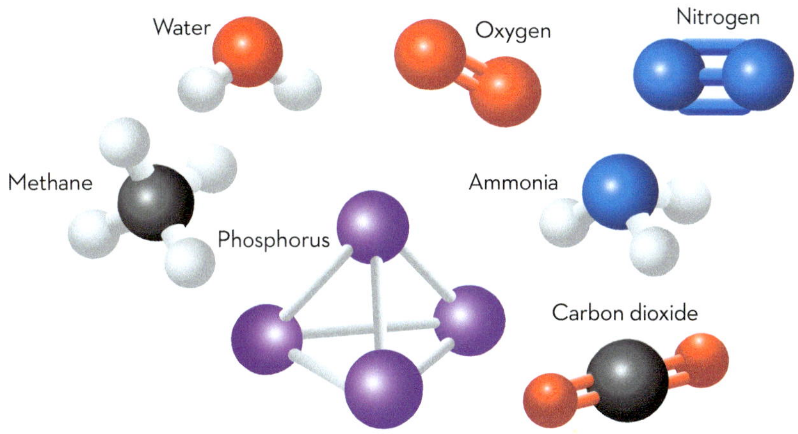

CHEMISTRY 2.0

The following website lets you visualise many molecules in 3D:

goo.gl/rUyDEk

CLASSROOM LABORATORY **Visualising DNA**

2. The DNA molecule, which is responsible for genetic codes, is a molecule made of other, different molecules. That is why it is called a **polymer.** To see it you will need: washing-up liquid, salt, distilled water and alcohol.

   Dissolve a spoonful of salt in half a litre of distilled water.

   Pour a little water into another glass and rinse your mouth with it. Then spit the water (which now contains cells from your mouth) back into the glass. Now add two spoonfuls of the saline solution into the glass with your saliva in it.

   Next, dissolve a little washing-up liquid into water in a third glass. Then add a little of this solution to the glass with your cells in it and stir. The soap breaks the cell membranes and releases your DNA.

   Now pour a little alcohol into the glass and wait for two minutes until the solution separates into two phases (alcoholic and aqueous). You will see some fine white threads appear in the alcoholic phase. This is your DNA.

   Search for the following information on the DNA molecule: what it contains, its physical and chemical properties, and its function in living organisms.

   Write the procedure for this experiment as a series of steps, explaining clearly what you did.

CHEMISTRY 2.0

Watch this video to see some ways in which metallic bonds are formed:

goo.gl/gRDUve

WORKSHEET

Download this unit's worksheet and do activities 21 and 22.

6.2. Crystals

Crystals are groups of atoms ordered in a geometric structure: a cube, a prism, a pyramid, etc.

Crystals contain an extremely high number of **unit cells,** the smallest block in the crystal. You can see this in salt, for example.

Crystals

Ionic

Metal element atoms with non-metal atoms

Common salt (Sodium chloride NaCl)

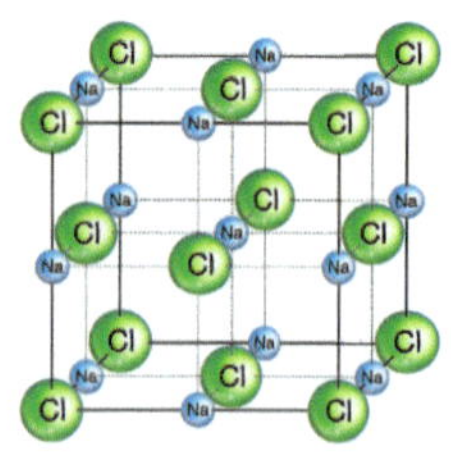

Plaster (Calcium sulphate $CaSO_4$)

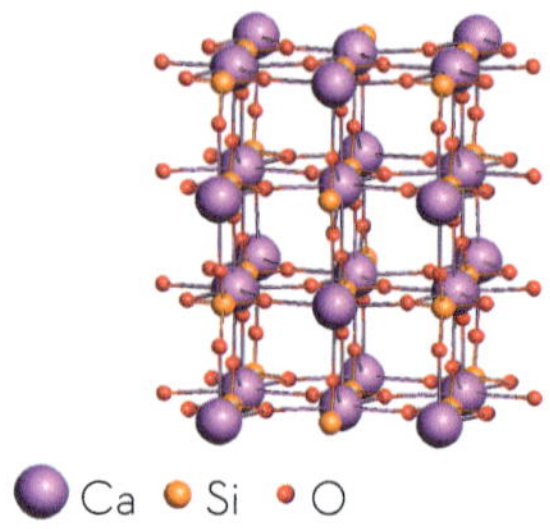

Covalent

Non-metal atoms with non-metal atoms

Quartz (Silicon dioxide SiO_2)

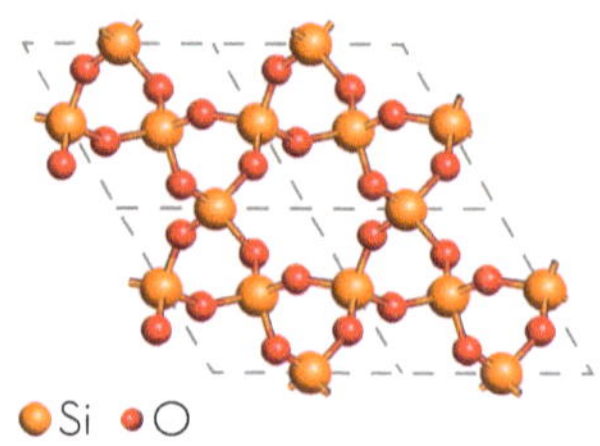

Diamond (C)

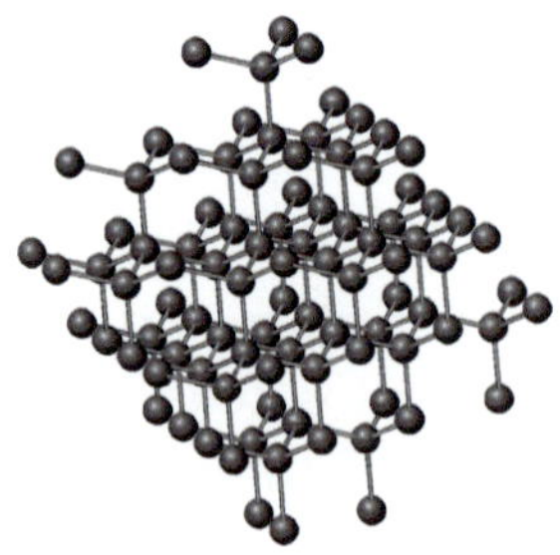

Metallic

Metal atoms with metal atoms

Iron (Fe)

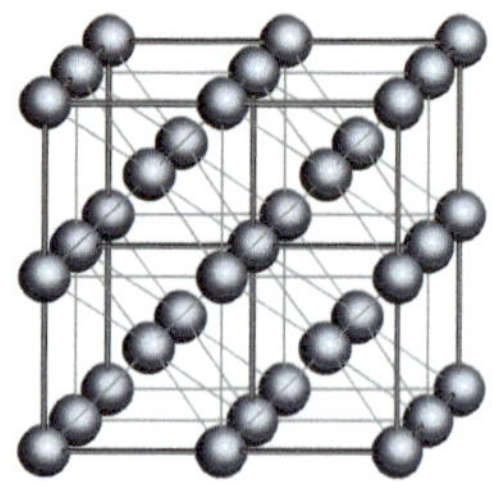

Copper (Cu)

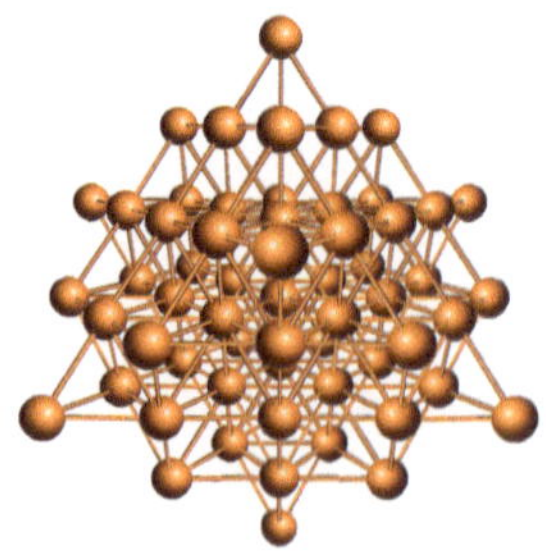

The properties of crystals can be summarised in the following table:

Crystals	Soluble in water	Melting temperature	Conductors of electricity
Covalent	No	Very high	No
Ionic	Yes	Very high	Only when dissolved or melted
Metallic	No	From low to high	Yes

ACTIVITIES

17. For the following molecular substances: H_2 (hydrogen), Cl_2 (chlorine), O_2 (oxygen), N_2 (nitrogen) and NH_3 (ammonia), indicate their physical state at room temperature and whether they are electrical conductors.

18. Indicate whether the substances made of the following atoms are molecules or crystals:

 a) Chlorine and calcium. *b)* Sulphur and hydrogen.

 c) Nitrogen and hydrogen. *d)* Oxygen and iron.

19. Copy the following chart into your notebook and complete it to determine the structure (ionic, molecular or crystalline covalent, or metallic) of a chemical substance:

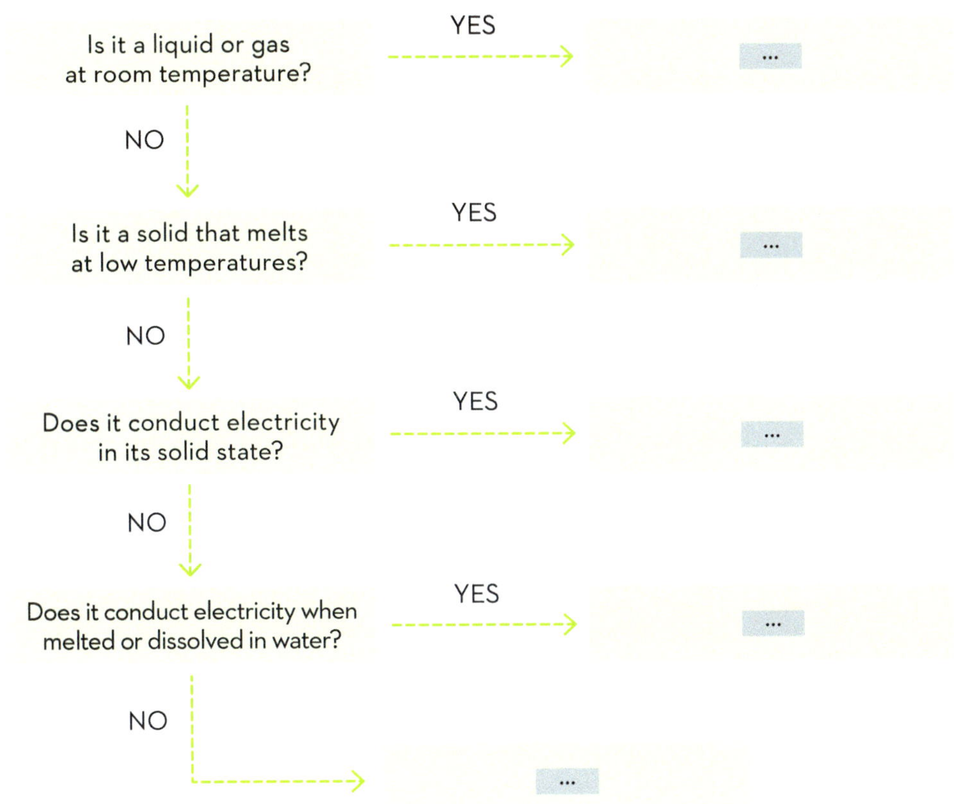

20. Match each of the following words to the correct sentence and compare your answers with a partner.

 Atoms Molecules Crystals

 a) Substances made only of non-metal atoms.

 b) They are always solid at room temperature.

 c) They constitute the inert gases.

 d) Made of a large number of atoms.

CHEMISTRY 2.0

A video explaining the formation of ionic bonds:

goo.gl/TwZjjX

MIND MAP

Download and print the worksheet you will find in the OLC to complete the mind map of the unit.

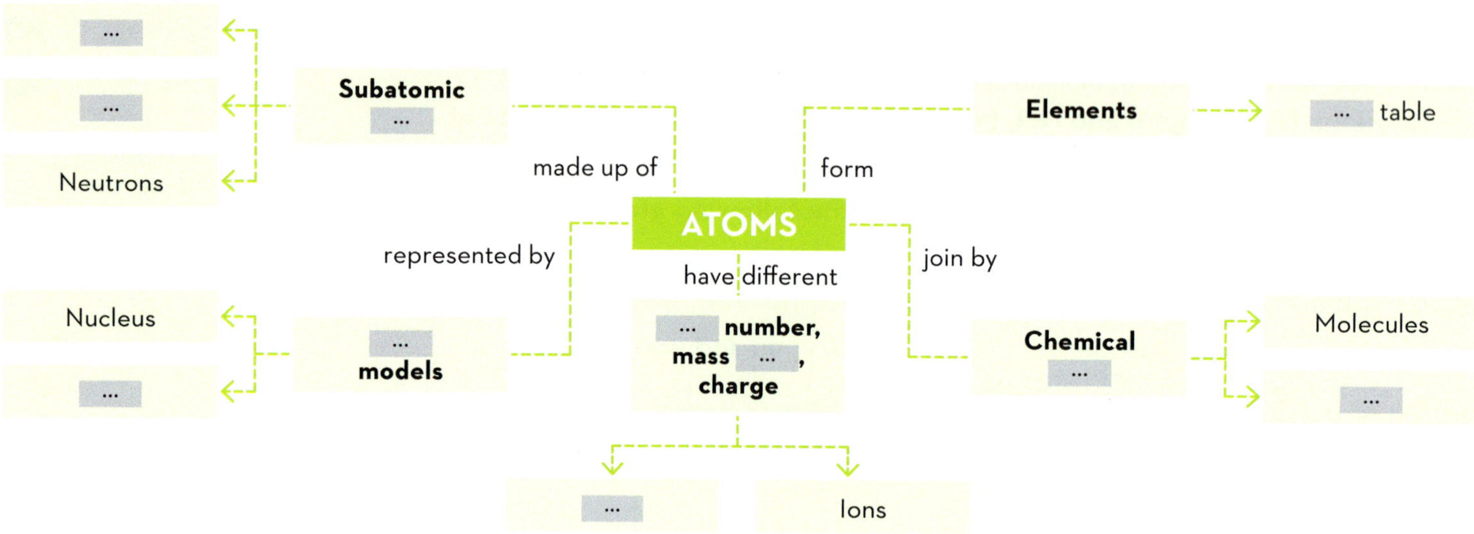

LOOK AROUND YOU

Raw materials

Even the most common everyday activities require a certain amount of raw materials. Material resources have always been essential, but the enormous consumption habits of modern civilisation have made them more important than ever.

Raw material is any basic and elemental substance used to manufacture products and obtain energy.

There are many types of raw materials and these are the basis for every industrial process. They can come from plants, animals (furs, hides, wool, fats, food, etc.) or minerals. The latter is the most abundant source and is extremely varied: fuels, metals, building materials, chemical products, water, etc.

Here are some of the main deposits:

- **Heavy metals:** present in the Earth's crust. Iron is the most important one (China).
- **Precious metals:** gold (South Africa)
- **Wood:** boreal (Canada and Russia) and tropical forests (Brazil).
- **Fuels. Natural gas:** northern Africa, North America, Central Asia. **Oil:** the Persian Gulf, the Caribbean, the Caspian Sea, Indonesia. **Coal:** South Africa, Central Europe, Central Asia, North America. **Uranium:** North America, Australia, South Africa.

Dept. of Education, Language Policy and Culture, Basque Government
www.hiru.com (adaptation)

Task

a) Play the Raw Materials Quiz at **goo.gl/ChiWqC.** Record your score and write down the new words you may find in the quiz.

b) Working in pairs, write a story in the first person about a natural resource of your choice, from the moment of its birth (the moment it was obtained) to its final destiny. You can write about where the story takes place, what you can hear, see and taste, how your journey goes, etc. Read your story to the rest of the class.

LABORATORY WORK

Chemical bonds based on the properties of substances

Objective

To test some properties (solubility, electrical conductivity and melting temperature) of various substances to find out what type of bonds they have.

Conductivity test circuit

Material

- Spatula.
- Crucible.
- Bunsen burner.
- Test tubes in a rack.
- Graphite electrodes.
- Battery and electrical connections.
- Small light bulb.
- Materials: copper sulphate, potassium chloride, pencil lead and sand.

Procedure

1. Solubility test

The crystals formed by the ionic bond of a metal and a non-metal are water-soluble. Put a small sample of the solid in a test tube, add a little water, shake it and observe whether it dissolves. Repeat with the other solvent (toluene). Write the results in the table below.

Safe method for heating a test tube

2. Electrical conductivity test

Ionic crystals do not conduct electricity when they are solids, but they do when dissolved in water. Covalent crystals, except graphite, never conduct electricity. Put a small solid sample in the crucible, introduce the electrodes, and close the circuit with the battery and the light bulb. Check whether the current passes through or not. Repeat the experiment with the solutions from the previous test.

3. Melting point test

Crystalline substances do not melt easily, but covalent molecules melt at low temperatures. Put a small sample of the solid in a test tube and carefully heat the bottom of the test tube over the Bunsen burner. Hold the test tube at an angle until the solid melts. If it melts at less than 100 °C, it will do so quickly, almost as soon as you start heating it. If it melts at between 100 °C and 300 °C, it will do so after a certain period of time.

Substance	Solubility		Conductivity		Melting point	
	In water	In toluene	Solid	Dissolved in water	< 100 °C	> 100 °C
Copper sulphate	...	...	...	...	...	...
Potassium chloride	...	...	...	...	...	...
Pencil lead	...	...	...	...	...	...
Sand	...	...	...	...	...	...

Analyse and respond

Compare the results you have collected about the characteristics you have studied for the three types of chemical bonds. Indicate and explain which of them will be present in each of the substances included in the experiment.

FINAL ACTIVITIES

Basic activities

1. Indicate which of the following are simple substances and which are compounds: oxygen, ammonia, ozone, mercury, 96% alcohol, oil, plastic, a lump of coal and pipe. Compare your classification with a partner.

2. Matter has electrical properties:
   a) Because it contains electrically charged particles.
   b) Because it contains electrons.
   c) Because it contains protons.
   d) It does not have electrical properties, because it does not contain charged particles.

3. Choose the correct definition of an atom:
   a) A positive sphere with negative charges encrusted in it.
   b) A negative sphere with positive charges encrusted in it.
   c) A positive nucleus with negative charges revolving around it.
   d) A negative nucleus with positive charges revolving around it.

4. Research as a small group about 'quarks'. Where can they be found?

5. Indicate which of the following contain atoms: a spark, a bacterium, a piece of wood, a flower, an echo, raindrops and a hair from your head.

6. If a neutral atom has 16 protons and 16 neutrons, how many electrons does it have? What is its atomic number? What is its mass number?

7. The element potassium (K) has Z = 19 and A = 39. Calculate how many protons, neutrons and electrons it has.

8. The element copper (Cu) has Z = 29 and 30 neutrons. Calculate how many protons and electrons it has. Calculate its atomic number (A). Compare your answers to 6, 7 and 8 with a partner.

9. Choose the correct definition of an isotope:
   a) Atoms of the same element with different numbers of electrons.
   b) Atoms of the same element with different numbers of protons.
   c) Atoms of the same element with different numbers of neutrons.
   d) Atoms of the same element with different numbers of these three particles.

10. Explain the difference between natural and artificial isotopes.

11. Work with a partner to search the internet for information on three artificial isotopes used in medicine and comment on their applications.

12. Choose the correct definition of an ion:
    a) Atoms which have gained or lost electrons.
    b) Atoms which have gained or lost protons.
    c) Atoms which have gained or lost neutrons.

13. Challenge a partner on the following elements represented by their symbols. Which one is different from its English name? N, Ne, He, Cu, Ni and Pt.

14. If you have a substance which is a good conductor of electricity in its solid state, is it a metal or a non-metal?

Consolidation activities

15. Look at the following atoms: $^{14}_{7}N$; $^{31}_{15}P$; $^{108}_{47}Ag$.

a) Indicate what each nucleus contains.

b) How many electrons does each one have?

16. Work with a partner. Copy and complete the following table in your notebook.

Atom	Z	A	Protons	Neutrons	Electrons
C	6	12	...	...	...
Zn	30	...	...	36	...
Hg	...	...	80	120	...
Br	...	80	...	...	35

17. An ion of an aluminium atom (Z = 13 A = 27) contains 10 electrons. Indicate the ion's charge and how many neutrons this atom has.

18. Which of the following atoms of elements are isotopes? Explain your answers.

$^{23}_{11}A$ $^{19}_{9}B$ $^{22}_{11}C$ $^{39}_{19}D$ $^{40}_{19}E$

19. Why do think ionic crystals can conduct electricity when they are dissolved or melted but not when they are solid? Sketch a simple diagram to illustrate your ideas.

Advanced activities

20. Explain the difference between an oxygen atom and the chemical element oxygen.

21. Work in groups of three. Search for information on the internet and use it to create a poster to explain how the isotope C-14 is used to determine the age of archaeological remains.

22. Find information to discover the physical properties and current main uses of the following elements: magnesium, copper and sulphur. Where would you put them in the periodic table: with the metals or the non-metals?

23. Use a computer to find the formulas for the following commonly used substances: isopropyl alcohol (ethanol), sugar, butane gas, chalk and bicarbonate of soda. Are the formulas the same in a Spanish website?

24. Indicate what type of bonds would form between the following pairs of elements:

a) S and Na. *b)* Cl and O. *c)* K and Fe.

d) F and Ca. *e)* H and Cl.

25. Indicate which of the following substances are found in nature, mainly as isolated atoms, molecules or crystals:

a) Diamond. *b)* Helium.

c) Nitrogen. *d)* Quartz.

e) Carbon dioxide. *f)* Gold.

g) Hydrogen. *h)* Common salt.

26. Using the data from the table below, indicate which substances are ionic and which are covalent:

Substance	Solubility in water	Electrical conductivity	Melting point
A	Yes	No	700 °C
B	No	No	−80 °C
C	No	No	1 250 °C

USE YOUR SKILLS

Acid rain

Below is a photo of statues called the Caryatids, built in the Acropolis in Greece over 2 500 years ago. The statues are made of a type of rock called marble. Marble is made of calcium carbonate.

In 1980 the original statues were moved inside the Acropolis Museum and replaced with copies. Acid rain was corroding the original statues.

Question 1

Normal rain is slightly acidic because it absorbs some of the carbon dioxide in the air. Acid rain is more acidic than normal rain because it has also absorbed other gases. These gases are sulphur dioxide and nitrogen oxide.

Where do the sulphur dioxide and nitrogen oxide in the air come from?

Question 2

We can simulate the effect of acid rain on marble by submerging marble chips in vinegar overnight. Vinegar and acid rain have almost the same level of acidity. Gas bubbles form on the marble chip when it is put in vinegar. We can measure the mass of the dry marble chip before and after the experiment.

A marble chip has a mass of 2 g before we submerge it in vinegar overnight. The next day, we remove and dry the chip. What is the mass of the dry marble chip now?

- Less than 2 g.
- Exactly 2 g.
- Between 2 and 2.4 g.
- More than 2.4 g.

Question 3

The students who performed this experiment also put marble chips in distilled water overnight. Explain why they included this step in their experiment.

Question 4

Write the procedure for your experiment as a series of steps.

The carbon footprint

Almost all the activities we do and all the material goods we use involve energy consumption. If this energy comes from fossil fuels (coal, gas or oil), it releases greenhouse gas emissions into the atmosphere. One of the main emissions is carbon dioxide (CO_2), although it is not the only one.

Climate change is essentially caused by the emission of greenhouse gases into the atmosphere as a result of socioeconomic activities.

The carbon footprint is the measurement of each individual's contribution to climate change. It calculates the emissions of CO_2 generated by our daily activities. In Spain 25% of all CO_2 emissions come from transport.

Objective

To calculate how much CO_2 is produced by different types of typical journeys.

Final product

Prepare a report on the main gases that cause the greenhouse effect and how we can reduce emissions. Indicate which other activities (apart from transport) generate the emission of CO_2 gases. Give ideas on how to reduce them.

Step by step

Observe the graph. It has a lot of information.

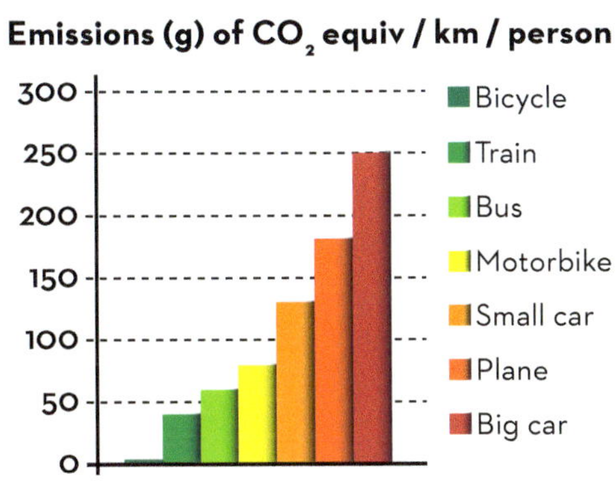

The vertical axis shows the equivalent emissions of CO_2 in grams per kilometre per person, from 0 to 300. The bars on the graph show the emissions caused by the different types of transport shown on the left axis.

Imagine you want to know how much CO_2 is emitted when you travel 5 km from your house to the football ground on your motorbike. The pollution for motorbikes is shown in the fourth column: 80 g per kilometre per person. If you travel alone for 5 km, it will produce 5 km · 80 g/km = 400 g of CO_2.

Now it is your turn. Calculate an estimation of how much CO_2 the following journeys emit:

a) Riding your bike to school.

b) Travelling by bus to the town centre.

c) Travelling by car with your family to a town less than 500 km from your town.

d) Travelling by plane with your family to a city of your choice in Europe.

Then prepare the report with a group of your classmates. You can even make a video or prepare a presentation. As a group, write a letter to your town/city mayor about CO_2 emissions and listing your ideas to reduce them.

Self-assessment

Evaluate your work and your classmates' work with the guidelines that you can download from the OLC.

5 CHEMICAL REACTIONS

Imagine a universe full of many different substances, but none of them can transform into another. Would there be any changes?

Changes mean that particles interact, or react, and they do this at a certain rate. We need to know how to control this rate.

For example, you have probably heard of nitroglycerine, an explosive substance. It is tremendously sensitive to impacts and handling it is very dangerous. This makes it almost impossible to use. However, a young chemist was able to convert it into another element that is more stable: dynamite. The name of the young man is probably familiar to you. His name was Alfred Nobel.

Chemistry is in our lives, whether we like it or not. Most of the time it helps us progress and live better, but it may also have a negative side. What would happen if the carton package, the plastic bottle or the aluminium foil covering the sandwich in your school bag ended up in the rubbish or on the ground and not in the appropriate bin?

If each student in your class produced one piece of rubbish during the break at school, there would be around thirty pieces only in that class, and between five hundred and one thousand in one school. That is forty-seven million in Spain and seven hundred forty-three million in Europe. In industrialised societies we generate one and a half billion pieces of waste in our break time. Don't you think something should change?

'The impossible of today will become the possible of tomorrow'.

Konstantin Tsiolkovsky (1857-1935), father of Russian cosmonautics

Contents

WORK WITH THE TEXT

These sentences are scrambled. Rearrange the words to create a true sentence, according to the text.

1. chemicals, in, made, Everything, is, the, Universe, of.
2. produce, useful, Chemical, substances, and, reactions, new.
3. any, change, reaction, substances, into, In, chemical, different, products.
4. the, 19th, In, chemists, invented, century, dyes, artificial.
5. may, negative, also, chemicals, a, Some, have, side.
6. environmental, away, packages, carelessly, Throwing, is, problem, an.
7. appropriate, rubbish, We, in, throw, the, away, bin, must.
8. waste, produced, of, Europe, are, Millions, of, tonnes, alone, in.

CHALLENGE

Chemistry is everywhere, and you can make simple chemical experiments at home. For example, place a hard-boiled egg (with the shell) in a jar and cover it with vinegar. Wait three days and write down your observations. Rinse the shell carefully with water and bounce it against a hard surface. What happens?

Write a short report on your experiment and share it with your classmates.

KEYWORDS

Download and print the activities that will guide you through the essential vocabulary of this unit.

FIND A CLASSMATE

Download and print the worksheet 'What do I want to know?'. Share experiences about this topic with your classmates and discover interesting things about them.

1 Changes in matter

If you look around, you will see that nature is constantly changing. But are all the changes the same? Using the first stage of the scientific method, **observation**, we will try to answer this question.

EXPERIMENT What happens in each case?

1. We propose five experiments to understand the transformations that can happen in nature. Write down all the steps and observations, especially the substances you start with and the ones you end with.

a) 

Colour explosion. Fill a plate with milk and add two drops of food colouring in various colours. Dip a cotton bud in washing-up liquid and then dip it in the centre of the plate.

b) **Ice melting.** Take an ice cube and put it in a glass. Let enough time pass until it has become liquid.

c) 

Lemon juice in milk. Squeeze half a lemon into a half a glass of milk. You can also mix cola and milk.

d) **Lemon juice and sodium bicarbonate.** Squeeze a little lemon juice into half a glass of water. Then add half a spoonful of sodium bicarbonate.

e) 

Glow stick. Buy a glow stick and break it or shake it (depending on the model) until it lights up.

Take photographs of the experiments and make a poster. Classify the experiences according to the appearance of new substances. Alternatively, you can record videos and make a video report.

Physical-chemical transformations can be classified as:

- **Chemical transformations:** Substances disappear and other new ones appear. These are called **chemical reactions.**
- **Physical transformations:** We have the same substances at the beginning and at the end of the process.

CHEMISTRY 2.0

You can see some interesting chemical reactions in the following links.

Colour explosion:

goo.gl/kamF3X

Homemade slime:

goo.gl/bSwiSG

Fluorescent slime:

goo.gl/t2QxEx

WORKSHEET

Download this unit's worksheet and do activity 1.

IMPORTANT

There are physical-chemical transformations that involve both types of changes.

Biochemistry is the branch of science that studies changes in living beings.

CLASSROOM LABORATORY **Experimenting...**

1. Until now you have studied clear, simple transformations, but sometimes they are not so evident. From the results obtained, try to find out what type of transformation occurs in these experiments.

1. The mysterious copper

Obtain a dilute solution of copper (II) sulphate.

- Write a word with the solution on a piece of filter paper and let it dry. Exchange it with a partner. You must wipe it with a piece of cotton dipped in ammonia to read what it says. You can also heat it carefully over a flame.
- Put a little solution in a test tube and add ammonia gradually with a dropper. Continue adding until you do not see any more changes.

Record your observations and write a brief description of the experiment.

2. The foam monster

Put 15 mL of hydrogen peroxide into a graduated cylinder. Next add 5 mL of washing-up liquid. Mix these substances. Prevent foam from coming out or you will have to start again. Put the test tube in a sink.

Weigh 5 grams of potassium iodide on a piece of paper. Fold the paper and pour the potassium iodide into the test tube. Immediately move away so the monster does not attack you. (Let the foam cool before cleaning it with a lot of water and gloves.) Do not touch the foam with your fingers!

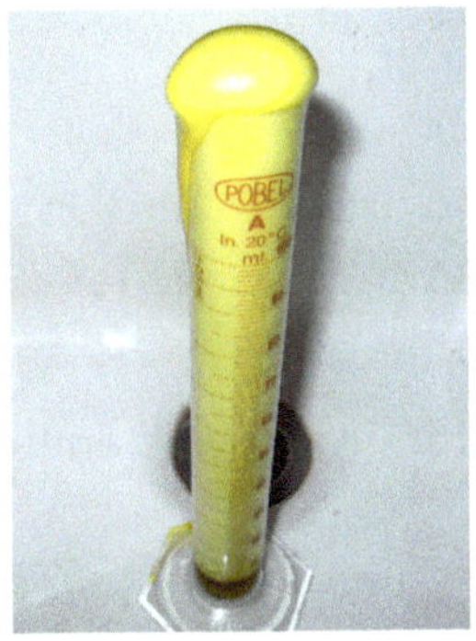

3. Purple vapours

Put some iodine crystals in a beaker and cover it with a watch glass. Heat the entire system in a fume cupboard or in a well-ventilated place (the gases are very harmful). When finished, look at the bottom side of the watch glass and wipe it with a paper tissue. Be careful because it will stain.

DID YOU KNOW?

The change of colour, the formation of a gas or the appearance of a precipitate (insoluble substance) can be signs of the existence of a chemical reaction.

However, we should always analyse the products, because its appearance may be a consequence of physical changes.

WORKSHEET

Download this unit's worksheet and do activities 2 and 3.

ACTIVITIES

1. **Look for information on the following processes and classify them as physical changes or chemical changes. Compare your answers with your partner.**

   a) A spark from a lighter.
   b) Silver darkens.
   c) A petrol engine.
   d) A fluorescent tube.
   e) Producing light with a dynamo.
   f) Effervescent medicine.

2. **Locate a transformation of each type around you. Do an experiment to test your hypothesis. Take photographs and make a brief presentation in class, showing your conclusions.**

2 Chemical reactions

As you have seen, the key aspect of a chemical change is the transformation of some substances into others.

A **chemical reaction** is the process by which some substances, called **reactants**, become new substances, called **products**.

Chemistry uses special notation to incorporate all this information:

A **chemical equation** is the representation of a chemical reaction:

Reactants — Direction of the process — Products

$$CH_4\,(g) + 2\,O_2\,(g) \longrightarrow CO_2\,(g) + 2\,H_2O\,(g)$$

The chemical formulas of the substances are indicated.

Coefficients indicate the proportion of the number of particles.

A letter identifies the state of the substance:
s: solid g: gas l: liquid aq: aqueous solution

WORKSHEET

Download this unit's worksheet and do activities 4 and 5.

In order to understand better what occurs, we will analyse the transformations at the atomic level. We will study two examples, observing the atomic structure of the substances involved.

Firstly, we will analyse the combustion of methane we saw above.

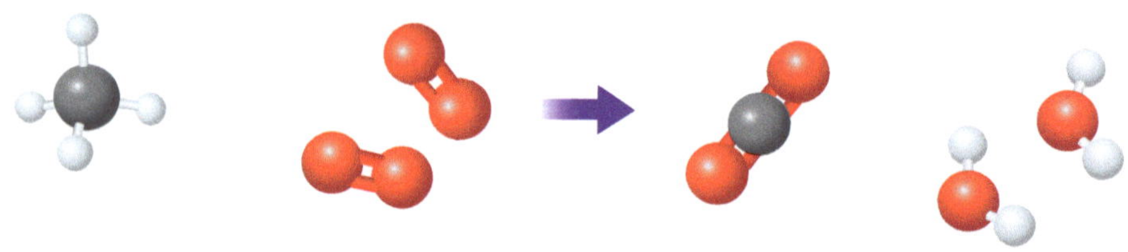

$$\mathbf{CH_4\,(g) + 2\,O_2\,(g) \rightarrow CO_2\,(g) + 2\,H_2O\,(g)}$$

The second example is the formation of ammonia from its elements.

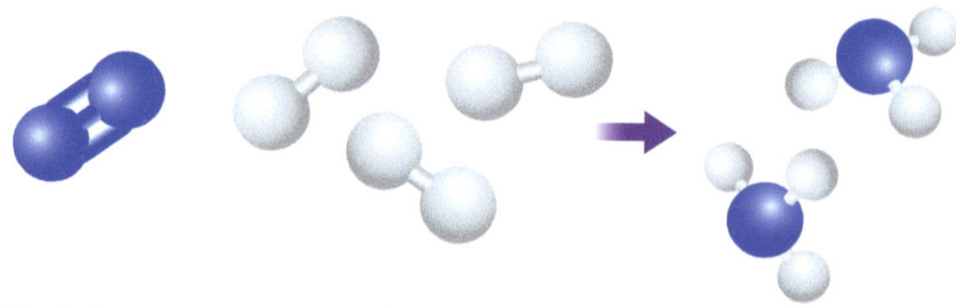

$$\mathbf{N_2\,(g) + 3\,H_2\,(g) \rightarrow 2\,NH_3\,(g)}$$

If we examine the representations above, we can conclude that:

- One or several substances, called **reactants**, react and transform.
- One or several substances, called **products**, are formed.
- No atom disappears, and no atom appears that was not there at the beginning. There has only been a **redistribution** of the atoms.

CHEMISTRY 2.0

This simple applet will review the concepts learnt in this section. Go to:

goo.gl/iC1XwX

CLASSROOM LABORATORY **Identifying products**

2. Take photographs of the following experiments and make a poster identifying reactants and products, and explain what occurred in each case.

> **WORKSHEET**
>
> Download this unit's worksheet and do activity 6.

1. Coppering is a gerund

Since the beginning of the 19th century, humans have been able to add a thin layer of a noble metal to worthless pieces to increase their beauty. Now you will do the same; you will **copper** something.

You need a piece of metal; it can be a spoon, a key or a clip. Make the assembly as indicated and let the electric current circulate for five to ten minutes. Weigh the object before starting the experiment and after finishing it. Why has the weight of the object increased?

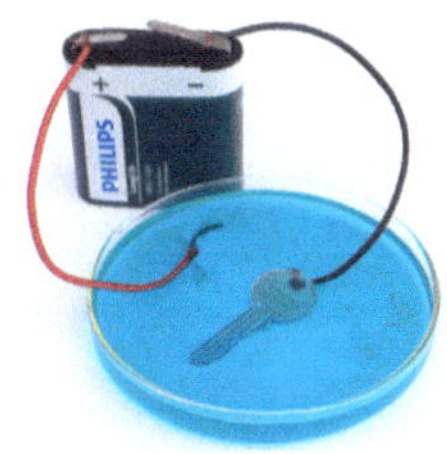

$Cu_2(SO_4)$ (aq) → Cu (s) + others

2. Detecting CO_2

Maybe you have thought about how we can be sure that we exhale $\mathbf{CO_2}$ when we breathe. We can check using lime water, a saturated aqueous solution of calcium hydroxide ($Ca(OH)_2$).

When carbon dioxide comes in contact with the solution, it reacts and precipitates as calcium carbonate according to the reaction.

$$Ca(OH)_2\ (aq) + CO_2\ (g) \rightarrow CaCO_3\ (s) \downarrow + H_2O\ (l)$$

Measure 50 mL of water and add a sufficient amount of calcium hydroxide ($Ca(OH)_2$) to saturate it. Then filter it to obtain a transparent, clean solution.

Blow gently through a straw into the solution. The appearance of a precipitate demonstrates the existence of CO_2 in the air we exhale.

Example video: **goo.gl/5vV7se**

3. Production of hydrogen

Hydrogen is a highly inflammable gaseous element that reacts violently with oxygen to create water. We produce it by making zinc (Zn) react with hydrochloric acid (HCl), although it also works with aluminium and cola, because this fizzy drink is acidic.

$$2\ HCl\ (aq) + Zn\ (s) \rightarrow ZnCl_2\ (aq) + H_2\ (g) \uparrow$$

Dissolve 10 cm^3 of hydrochloric acid in 50 mL of water in an Erlenmeyer flask. Add a little zinc to the mixture. Cover it and collect the gas above a crystalliser with water and washing-up liquid as shown in the picture.

We can easily identify the gas by taking a spoonful of the bubbles and lighting a match to it. The hydrogen burns producing a little 'pop' explosion:

$$2\ H_2\ (g) + O_2\ (g) \rightarrow 2\ H_2O\ (g)$$

Example video: **goo.gl/U57CxT**

3 Law of conservation of mass

CHEMISTRY 2.0

You can determine reactants, products and leftovers in the following applet:

goo.gl/Z15kV

WORKSHEET

Download this unit's worksheet and do activities 7, 8 and 9.

So far you have made some chemical transformations and we have tried some processes to identify products. But what happens to the **mass** in a chemical reaction?

This is the question Antoine Lavoisier asked himself in the middle of the 18th century. In order to solve this, he began to experiment with closed containers, carefully weighing reactants and products. In the following section you will use the same technique as Lavoisier, but you know about something that he did not: the existence of atoms. Now we will answer his question, using the reaction of the combustion of chlorine:

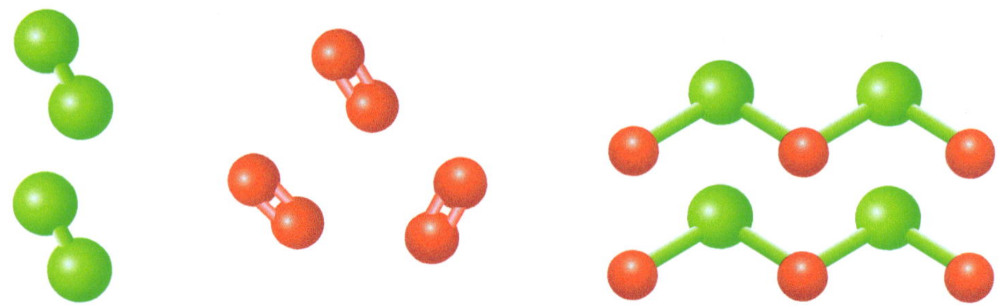

$$2\,Cl_2\,(g) + 3\,O_2\,(g) \rightarrow 2\,Cl_2O_3\,(s)$$

Look at the reaction above. Has any atom appeared or disappeared?

Using the atomic masses, determine the mass of all the reactants and compare it with that of the products. Did it vary during the process?

As each element has an **atomic mass**, we can deduce:

In a chemical reaction, no atom disappears or appears, so the mass of the reactants is equal to the mass of the products. **Mass is conserved.**

ACTIVITIES

3. By applying the Law of Conservation of Mass, predict the mass of the other compound:

	HCl	+ NaOH	→ NaCl	+ H_2O
a)	73 g	80 g	117 g	?
b)	?	20 g	29.25 g	9 g

	N_2	+ 3 H_2	→ 2 NH_3
c)	56 g	?	68 g
d)	140 g	30 g	?

4. When oxidising magnesium according to the reaction 2 Mg + O_2 → 2 MgO, you wrote the masses of the solids, 72.9 g and 120.9 g, but you do not know which corresponds to each solid. Assign them correctly and explain your answer. Compare your answer with a partner.

5. When you open a fizzy drink, does its mass increase or decrease? State your hypothesis and debate in a small group.

6. Which has a larger mass, a new iron nail or a rusty iron nail? Explain the reasons for your answer and draw the reaction with atoms.

We must give credit to the effort Lavoisier made to establish the Law of Conservation of Mass. Today we have the concept of atoms, but in that time period Dalton had not formulated his atomic theory.

CLASSROOM LABORATORY Checking the Law of Conservation of Mass (LCM)

3. We suggest you step into Lavoisier's shoes and try to observe his contribution through your experiments. Remember that the use of the scale is fundamental. Keep in mind that the main contribution of the scientist was due to the use of this instrument.

Stage I

- Weigh 1 gram of sodium bicarbonate in a watch glass.
- Measure 5 cm^3 of vinegar.
- Weigh everything.

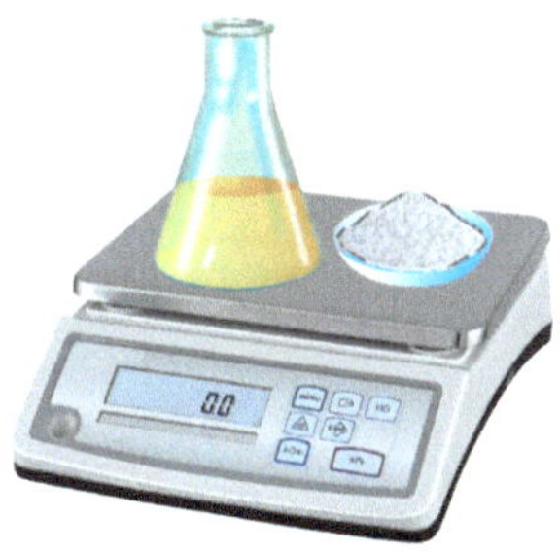

- Pour the solid onto the vinegar.
- Weigh everything again (including the empty watch glass).

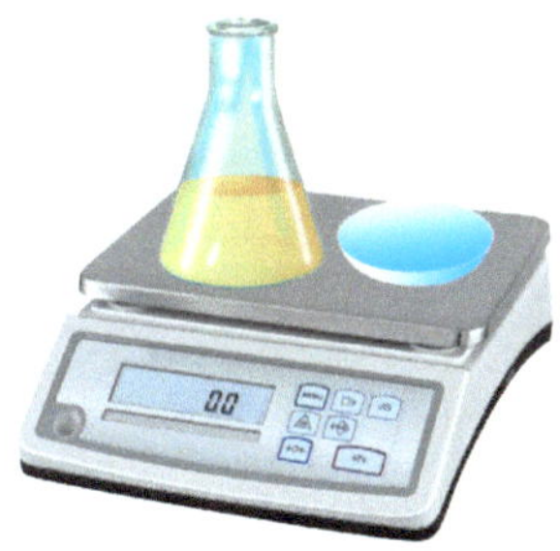

a) What is the concentration of sodium bicarbonate in g/cm^3 and in kg/L?

b) Was the mass conserved in the process?

c) Observe what happened during the reaction and present a hypothesis to explain the result.

Stage II

Repeat the experiment, but this time, use a balloon to contain the sodium bicarbonate and pour it into the Erlenmeyer flask.

a) What can you deduce from the result of the experiment?

ACTIVITIES

7. You will test which combinations there should be between the reactants and the products for the Law of Conservation of Mass to be fulfilled. Choose the version of the application that interests you most and start with the introduction.

In tools, activate the scale option and try to adjust the quantity of atoms you use to balance the scales. You can change the reaction and hide the particles.

After this, try the game. Take a screenshot of each example you solve and make a presentation explaining how you did it and showing your achievements. Write a procedure indicating which elements you balanced first. Use connection words such as firstly, then, finally, etc.

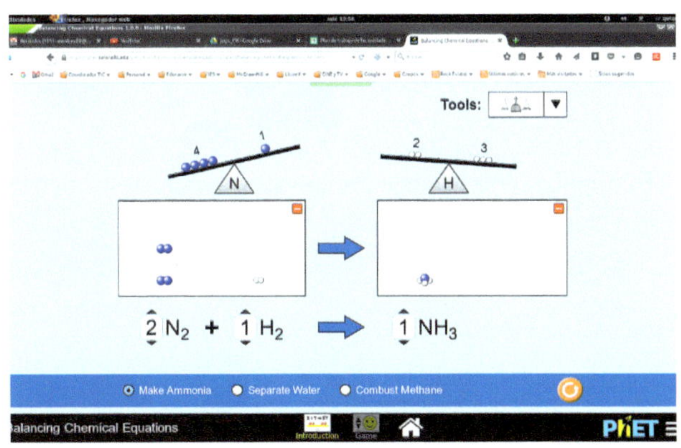

Go to the simulation at: **goo.gl/bzLBov**

4 Energy in reactions

Chemical reactions change substances into others, but we can also obtain other benefits from them.

THINK AND REASON

In the reactions of the section Classroom Laboratory 2, did you observe other physical effects such as heat, light or electricity?

Think about the possible applications of the previous reactions, according to the effects obtained.

The recombination of atoms in a chemical reaction means that the initial bonds break and new bonds form.

$$2\,H_2O\,(g) \rightarrow 2\,H_2\,(g) + O_2\,(g)$$

Some of the atoms that form the **reactants** must separate (break bonds). This requires an amount of energy.

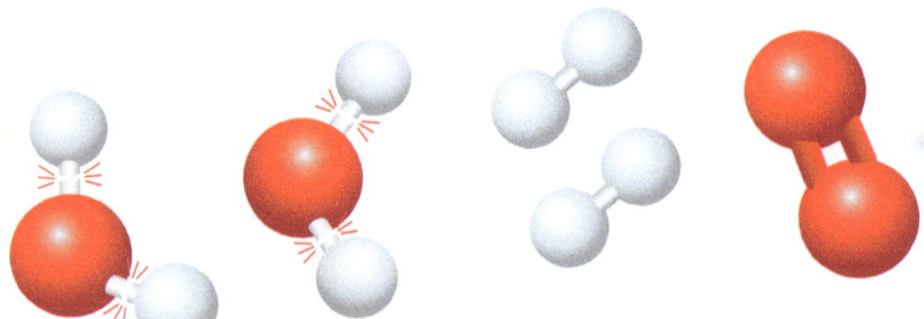

Later, when they join again to form the **product** (form bonds), they release another amount of energy, which is different from the previous energy.

By comparing both amounts of energy in what is called the **energy balance**, we can have two situations:

Exothermic processes

The **energy** needed to separate the atoms (break the bonds) is **lower** than the energy released when they join again (form new bonds).

$4\,Na\,(s) + O_2\,(g) \longrightarrow 2\,Na_2O\,(s)$ Oxidation of sodium

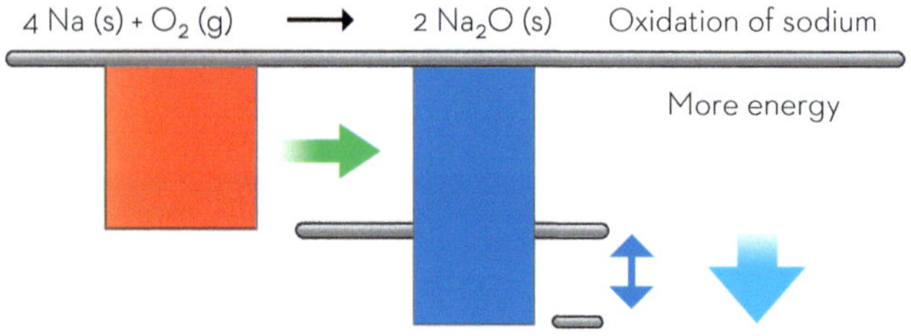

Examples: combustion, respiration, etc.

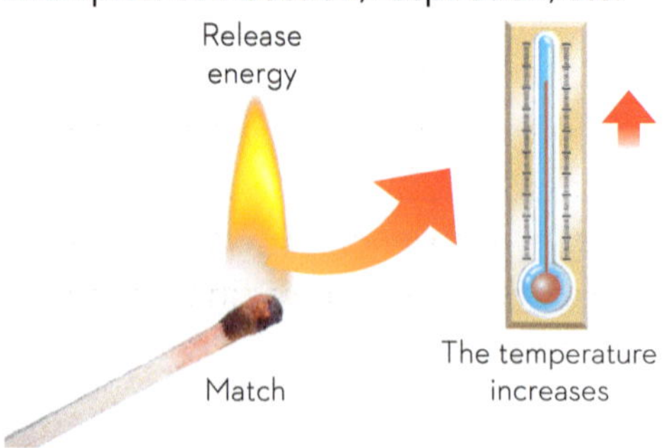

Endothermic processes

The **energy** needed to separate the atoms (break the bonds) is **greater** than the energy released when they join again (form new bonds).

$3\,O_2\,(g) \longrightarrow 2\,O_3\,(g)$ Formation of ozone

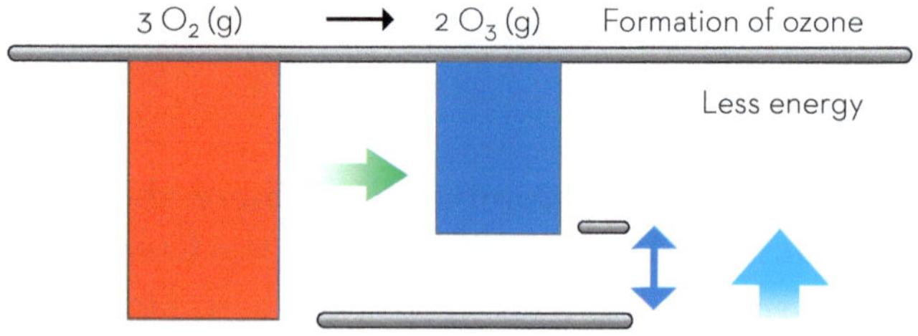

Examples: dissolving potassium nitrate, photosynthesis, etc.

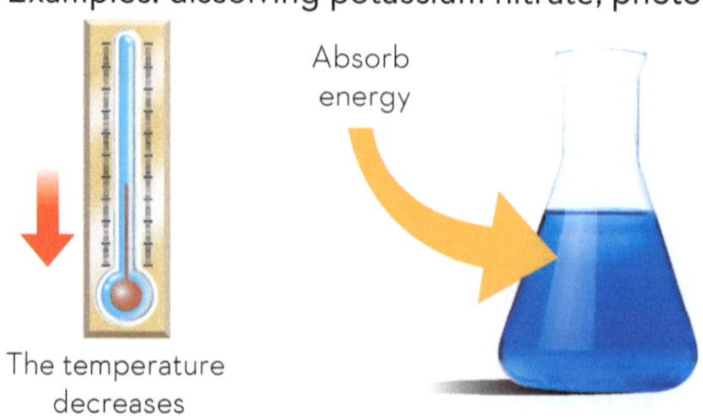

CLASSROOM LABORATORY **Playing with energy**

4. This time we suggest you do two controlled experiments that will help you understand energy balance better. As a final product, we recommend that you make a video report in groups in which you explain what happened with the temperature and why.

1. Exothermic reaction

We suggest you simulate the reaction used in self-heating cans for some prepared drinks like coffee.

Take a beaker with 50 mL of water and write down its temperature. Next, add 20 g of anhydrous calcium chloride ($CaCl_2$) and stir gently. Measure the final temperature to determine how much it has increased. Be careful because the beaker may be very hot. Put it on a stable, safe base.

2. Endothermic reaction

These reactions are not very common because they need an external source of energy. If no energy is provided the process will stop. However, sometimes the process is so favourable that it directly absorbs the energy needed from its surroundings, provoking a decrease in the temperature.

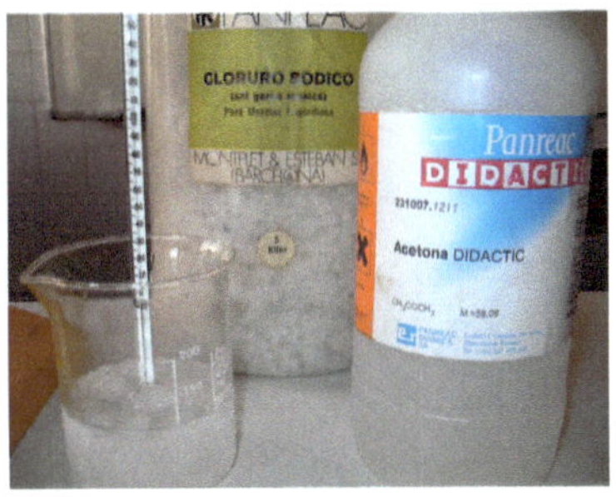

Some solutions are examples of this type of behaviour. In the following experiments, compare the temperature before and after the mixing.

- One gram of potassium nitrate in 50 mL of water.
- Some crushed ice and salt.
- Some crushed ice and 100 mL of acetone.

VOCABULARY

Reactions where energy is released are called **exothermic** (exo-, Greek prefix meaning 'outside'). If energy is absorbed, they are called **endothermic** (endo-, Greek prefix meaning 'inside').

WORKSHEET

Download this unit's worksheet and do activities 10, 11 and 12.

IMPORTANT

Frostbite is the effect of exposure to very low temperatures. Its effects on the skin are similar to burns.

ACTIVITIES

8. **If process A → B is exothermic, what is the opposite, B → A? Explain your answer with a diagram.**

9. **We know sodium bicarbonate as baking soda. At high temperatures it decomposes, producing carbon dioxide. This makes bubbles in the mix. Is this decomposition an endothermic or exothermic process?**

10. **Put some food waste or leaves in an airtight plastic bag and close it tightly for two days. How can you analyse whether the fermentation produced is an endothermic or exothermic process? Write a short report on your experiment, indicating the procedure and conclusions.**

5 Reaction rate

You can identify a large number of chemical transformations around you.

- Some of them are very fast, like the detonation of a firecracker. We say that its **kinetics** is **fast**.
- On the other hand, iron oxidation and the deterioration of paper happen slowly. We say that their **kinetics** is **slow.**

The rate at which a process occurs depends on many factors. Let's look at two of the most important ones.

VOCABULARY

Kinetics: branch of science that studies the speed at which changes occur.

WORKSHEET

Download this unit's worksheet and do activities 13 and 14.

Water molecule

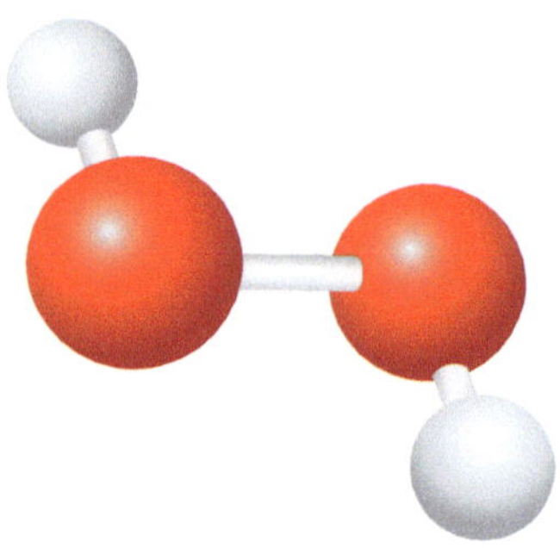

Hydrogen peroxide molecule

5.1. Influence of temperature

It is one of the most important factors that affect the reaction rate.

THINK AND REASON

You have already studied the Kinetic Theory of Matter. Considering that chemical reactions are due to interactions between the particles of the reactants and that these must come close to each other, answer the following:

- How is the speed of the particles related to the temperature?
- How will the temperature affect the reaction rate?

The **higher** the temperature of a process, the **faster** the speed at which it occurs, because the particles move faster and interact more.

EXPERIMENT Decomposing hydrogen peroxide

2. Hydrogen peroxide is an unstable compound that decomposes slowly into water and oxygen. Take two beakers and pour 20 cm^3 of hydrogen peroxide in them. Slowly heat one of them. What do you observe?

ACTIVITIES

11. Look up information on baking powder. How does it work? What is its purpose?

12. You know that in a pressure cooker food cooks quicker than in an open pot. Link this increase in speed to what happens to the temperature and pressure inside a heated, closed pot. Can you imagine a condition or place where food cooks more slowly?

13. Can there be chemistry, and therefore, life on a world that is frozen or too hot? Explain your answer. Discuss in small groups.

14. It is common to keep food in the refrigerator or even in the freezer. Why do you think this is? What does it mean when frozen food packages say 'do not break the cold chain'? What does the number of stars on the freezer mean? Write your answers as a short essay. Share in small groups and decide on a final short report.

5.2. Influence of concentration

You have probably observed that natural (100%) orange juice has a more intense orange flavour than orange fizzy drinks (5-10% orange juice). If concentration measures the amount of substance present per unit of volume, will it have an influence on the reaction rate?

THINK AND REASON

When an irritating substance or toxic material touches the skin or eyes, why must we pour a lot of water on the affected area?

Look at the picture on the right. In which jar do you think there are more particle collisions, and therefore a higher reaction rate?

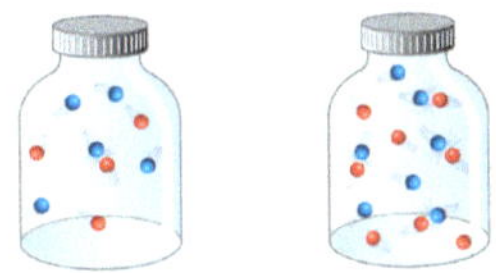

The **higher** the concentration of a reactant, the **higher** the reaction rate, because there is a larger quantity of particles available to react and they are closer together.

EXPERIMENT Eliminating colours

3. Take two containers with the same amount of bleach (be careful while handling) and add two glasses of water to one of them. Then place a coloured piece of cloth, preferably cotton (not synthetic) in both of them. You can substitute the cloth with the leaf of a chard or any other plant.

   Determine how much time the discolouration takes in each of the solutions. Make a video report to show what happened and give your explanation.

ACTIVITIES

15. **Working in pairs, investigate the relationship between dynamite and nitroglycerine and write a brief report, explaining the effect of concentration on this invention.**

16. **You have probably observed that after keeping hydrogen peroxide in the first aid kit for a year, it is not as effective. Why is this?**

DID YOU KNOW?

When you see a car exploding in a movie, it is fake. Fuel tanks do not explode because the concentration of oxygen is too low. It is necessary to have a mixture of fuel and air in the adequate proportions.

An excess of any of the reactants means there is a lack of the other and therefore the reaction will not occur.

WORKSHEET

Download this unit's worksheet and do activity 15.

6 Important chemical reactions

We can find different types of chemical reactions in nature. We will study some of the most important ones:

6.1. Acid-base reactions

Many substances around us can be classified as acids or bases. These react with each other according to the following reaction:

Acid + Base → Salt + Water

We measure acidity-basicity with the pH scale.
We use indicator paper to measure the pH. This changes colour according to the pH.

Acids: substances that have a sour taste and react with limestones such as marble.

Bases or alkalis: substances that have a bitter taste and soapy feel and do not usually attack metals.

EXPERIMENT The power of vinegar

4. Because vinegar is acidic, it reacts with bases. Observe the results in the following three examples, explain what occurred and make an explanatory leaflet with photographs.

Attacking lime scale	Producing gas	Dissolving hard things
Take an object that has lime scale, like a tap aerator, and submerge it in a glass of vinegar for two hours.	Add a spoonful of sodium bicarbonate to some vinegar.	Put a hard-boiled egg or a bone in a glass of vinegar for four hours or more.

WORKSHEET

Download this unit's worksheet and do activities 16 and 17.

ACTIVITIES

17. Research why we take sodium bicarbonate or magnesium chloride when we have heartburn. Work in small groups and write a short essay.

18. Some ants inject formic acid when they bite. How would you relieve one of these bites?

19. Indicate what colour the indicator paper would be with grapefruit, shampoo and cola fizzy drink. Check your answers. Compare your answers with your partner.

6.2. Chemical Reduction-Oxidation reactions (REDOX)

These are reactions where there is an exchange of electrons between elements:

- One element loses electrons (oxidation).
- One element gains electrons (reduction).

The electrons that one element loses are gained by the other element.

Some examples of **redox** reactions are:

- **Combustions:** one substance oxidises with the oxygen in the air.
- **Metal corrosions:** iron oxidation, silver tarnishing and aluminium losing its shine.
- **Batteries:** obtaining electricity from a reaction.

WORKSHEET

Download this unit's worksheet and do activity 18.

EXPERIMENT **Use fruit if you haven't got a battery**

5. At one point you may have needed a battery to turn a light bulb on. If this happens again, take a copper wire, a piece of aluminium and make the following assembly.

Now you have your battery. It will produce a voltage around 2 volts and it will last until the juice is gone.

You should sand the ends a little.

Use an orange, lemon, peach or any other fruit that has a lot of juice. You can also use a potato if necessary.

6.3. Other common reactions

- **Precipitation:** reaction that produces an insoluble product.
- **Fermentation and respiration:** biological process where energy is obtained from oxidation.
  - The respiration process starts with glucose and oxygen, and CO_2 and water are obtained.

$$\text{Glucose} + O_2\,(g) \rightarrow CO_2\,(g) + H_2O\,(l)$$

  - Fermentation is an incomplete oxidation. It does not use oxygen, and alcohols are the product, as in wine fermentation.
- **Decomposition:** a complex compound separates in two simpler substances.

$$CaCO_3\,(s) \rightarrow CaO\,(s) + CO_2\,(g)$$

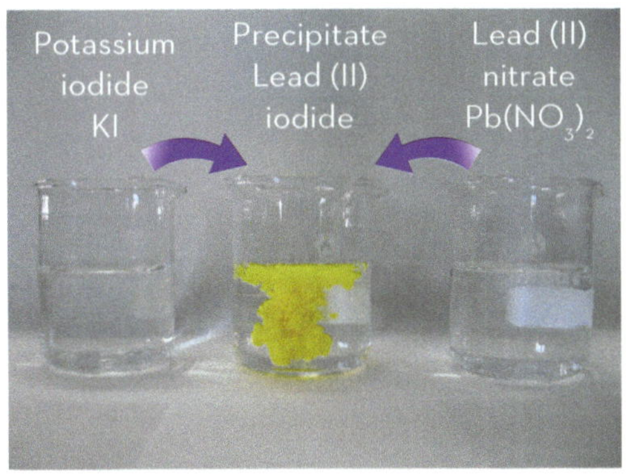

ACTIVITIES

20. What process is used to make wine and beer? Search for information about the process and make a brief presentation for the class.

21. If you were on the Moon, outside the spaceship, could you light a match? Discuss your answer with your partner.

7 Chemistry in our lives

750,000 years ago humans learnt to use chemistry: they found out how to light a **fire**. Since then, chemistry has changed our lives, although now there are few natural materials around us, like wood, cotton and linen. Can you think of any others?

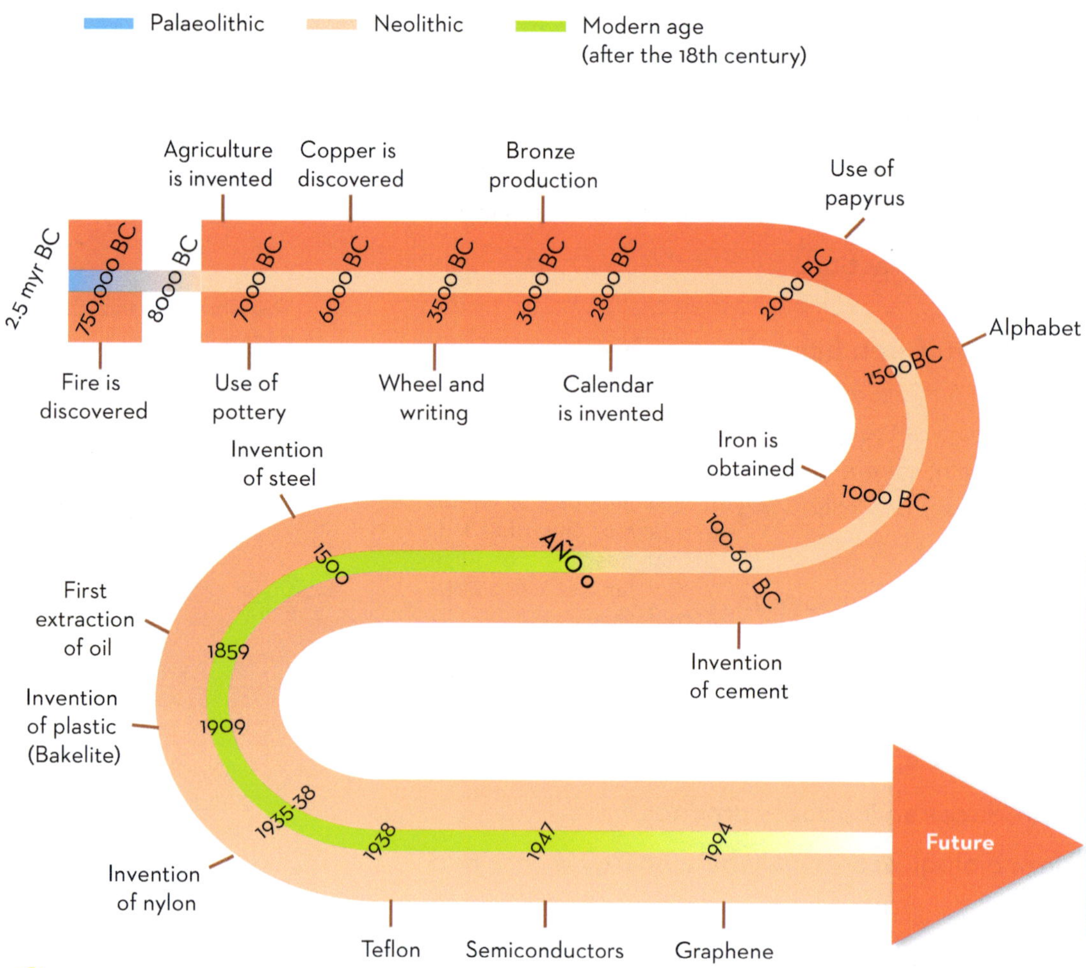

Materials with unexpected properties and the use of the 3D printer

Graphene aerogel

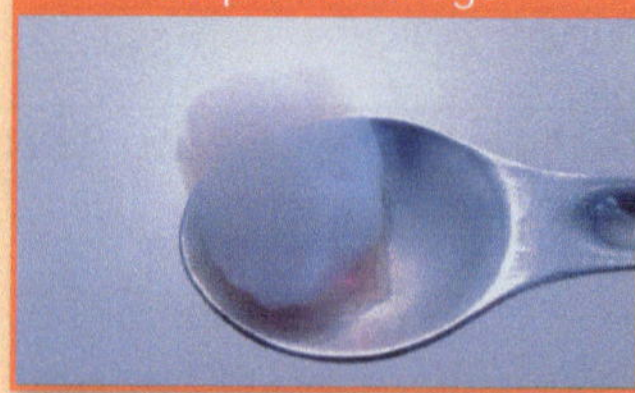

Graphene

Nanoparticles

3D printer

ACTIVITIES

22. Go to the following link: goo.gl/G9f3uh. Click on the *introduction* and find out which are the main types of products produced by the chemical industry. Describe them briefly. Compare your list with your partner.

23. Go to the following link: goo.gl/wE4fRk. Click on *basic chemicals* and look at the world production of the indicated products. Working in pairs, make a bar graph showing the ten most manufactured and the percent of the total.

24. Do a debate on the benefits and disadvantages that chemistry brings to our way of life. Divide the class into groups. Each one has ten minutes to prepare a rebuttal for each idea. The groups will face each other two groups at a time and they will defend a position chosen at random. The rest of the class will act as the jury.

25. Identify five materials around you that are not found in nature. Find out when they were invented. Share your list with the whole classroom so you all can make a timeline including everyone's contributions.

7.1. The origin of substances

You have probably heard or read phrases such as 'no chemical additives' or 'no chemical products'. In the previous unit you learnt that all matter in the universe is made from the atoms of about one hundred elements. Therefore, all the substances and materials we use and consume are 'chemicals'.

However, we can differentiate between materials according to their origin:

- **Natural:** obtained directly from nature, like wood, cotton, rock and clay.
- **Artificial:** obtained from a technological process. There are two possibilities:
  - **Manufactured:** through modification of a natural product, like obtaining steel from iron, oil from olives and fabric from cotton, silk or linen.
  - **Synthetic:** from manufactured materials, like vulcanised rubber, or completely new, like plastic and synthetic fibres.

Currently, the synthesis of new substances is one of the most important parts of chemistry. New substances are constantly appearing with properties that are more special and with applications that were unthinkable before.

DID YOU KNOW?

The animist theory stated that matter in living beings had a special 'vital energy'. This idea is wrong: for example, there is no difference between natural and synthetic sugar.

7.2. Identifying substances

Our lives without chemistry would be very different.

The pictures show the composition of some common materials around us and the situation a century ago. Can you see the difference?

ACTIVITIES

26. Identify all the materials in the picture of the modern kitchen in section 7.2 and classify them as natural or artificial. Make a table and compare with your partner.

27. Choose an environment: your house, the school or a car. Take a photograph and identify all the products obtained from the chemical industry and look for a natural alternative. How would this change its use? Exchange your list with your partner and check the differences and similarities.

WORKSHEET

Download this unit's worksheet and do activities 19, 20 and 21.

8 Chemistry and the environment

WORKSHEET

Download this unit's worksheet and do activities 22, 23 and 24.

Our progress is linked to the ways we obtain energy and the materials we use. We name historical time periods according to the predominant material at the time: the Stone Age, the Bronze Age and the Iron Age. We even call the present time period the Plastic Age.

The chemical industry gives us an infinite number of advantages by providing new materials with more adequate and innovative properties. But it also has certain negative aspects. Until recently, we did not consider these negative aspects.

Acid rain

Origin

The fuels used for locomotion and electricity generation emit sulphur and nitrogen oxides. These chemicals react with water in the clouds and produce acids.

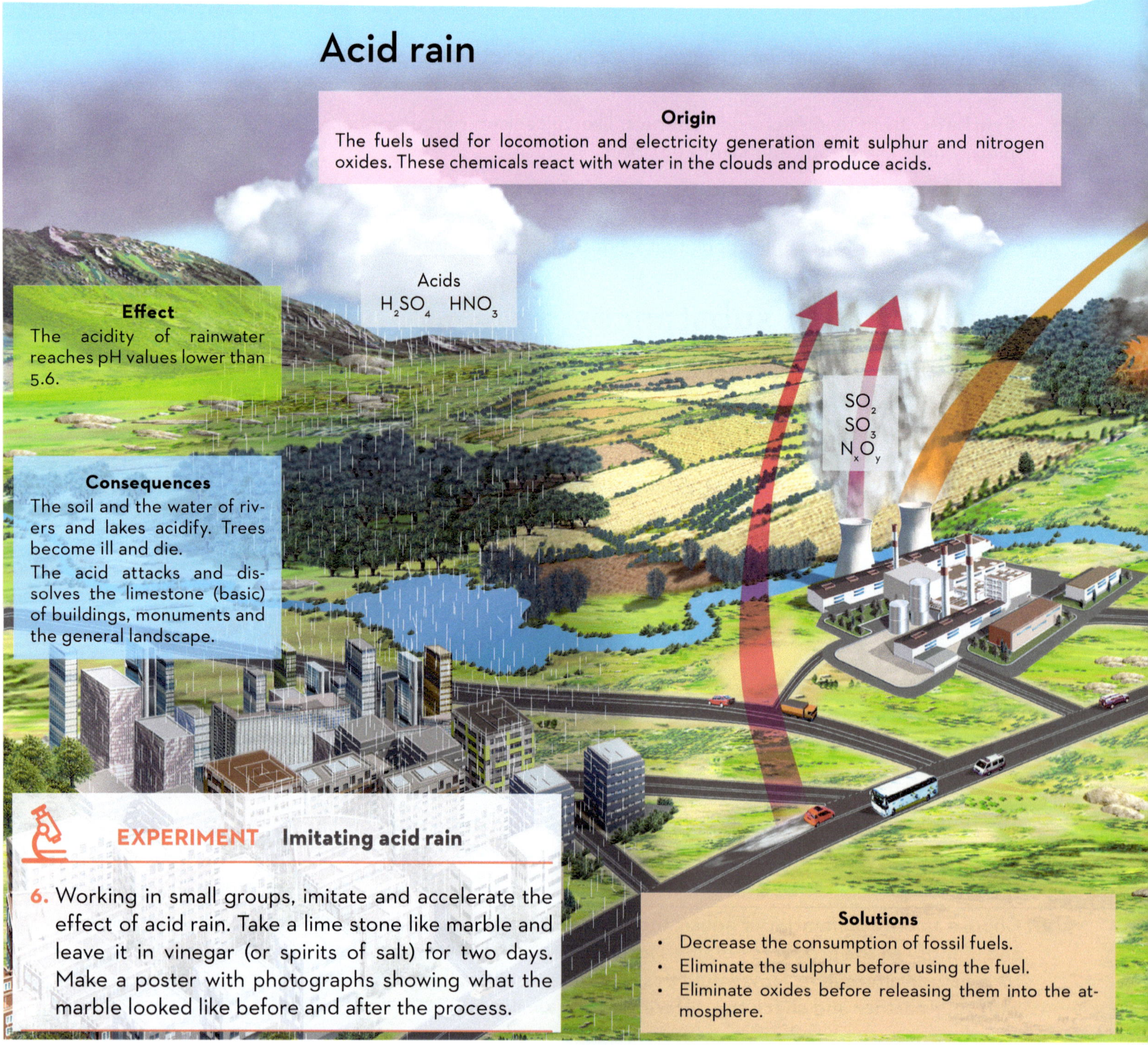

Effect

The acidity of rainwater reaches pH values lower than 5.6.

Consequences

The soil and the water of rivers and lakes acidify. Trees become ill and die.

The acid attacks and dissolves the limestone (basic) of buildings, monuments and the general landscape.

EXPERIMENT **Imitating acid rain**

6. Working in small groups, imitate and accelerate the effect of acid rain. Take a lime stone like marble and leave it in vinegar (or spirits of salt) for two days. Make a poster with photographs showing what the marble looked like before and after the process.

Solutions

- Decrease the consumption of fossil fuels.
- Eliminate the sulphur before using the fuel.
- Eliminate oxides before releasing them into the atmosphere.

The greenhouse effect

History

The Earth has an average temperature of 14.3 °C due to the presence of carbon dioxide (CO_2), methane (CH_4) and water vapour in the atmosphere. These gases prevent the planet's heat emissions from reaching space in the form of radiation. This causes a general warming effect. Without this effect, our planet would have a temperature of –19 °C.

Origin

The problem appears when the presence of CO_2, CH_4 and H_2O in the atmosphere is larger than normal. This produces an 'extra' greenhouse effect.

Effect

In recent years, we have detected an increase of this effect, caused by the increase in the concentration of these gases in the atmosphere.

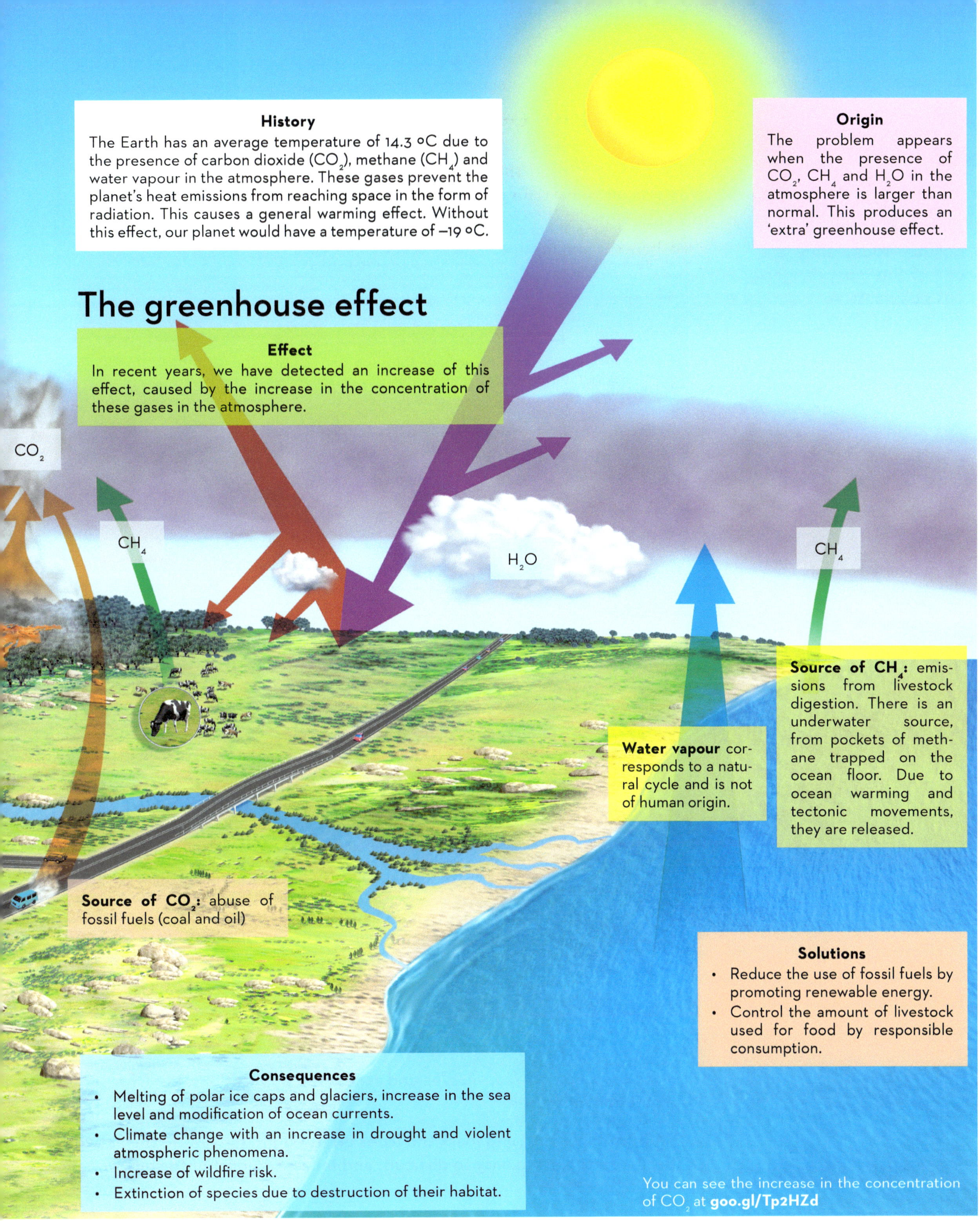

Source of CO_2: abuse of fossil fuels (coal and oil)

Water vapour corresponds to a natural cycle and is not of human origin.

Source of CH_4: emissions from livestock digestion. There is an underwater source, from pockets of methane trapped on the ocean floor. Due to ocean warming and tectonic movements, they are released.

Solutions

- Reduce the use of fossil fuels by promoting renewable energy.
- Control the amount of livestock used for food by responsible consumption.

Consequences

- Melting of polar ice caps and glaciers, increase in the sea level and modification of ocean currents.
- Climate change with an increase in drought and violent atmospheric phenomena.
- Increase of wildfire risk.
- Extinction of species due to destruction of their habitat.

You can see the increase in the concentration of CO_2 at **goo.gl/Tp2HZd**

Chlorofluorocarbons (CFCs)

DID YOU KNOW?

CFC molecules were designed explicitly not to pollute. Today they are one of the most dangerous pollutants.

History

In the decade of the 1950s an enormous hole in the ozone (O_3) layer was discovered. This is a thin layer located at 25 km of altitude. It has an elevated concentration of this gas. It is important because of the protection it provides to all life forms against the effect of ultraviolet radiation.

Effect

CFCs are inert in the lower layers of the atmosphere, but become very reactive in the higher layers when they come in contact with UV rays. One single molecule of CFC is able to cause 100,000 ozone molecules to break apart.

Origin

The cause of the hole in the ozone layer is due to chemicals called chlorofluorocarbons (CFCs). These are widely used as aerosol propellants and as refrigerants in refrigerators and air conditioning.

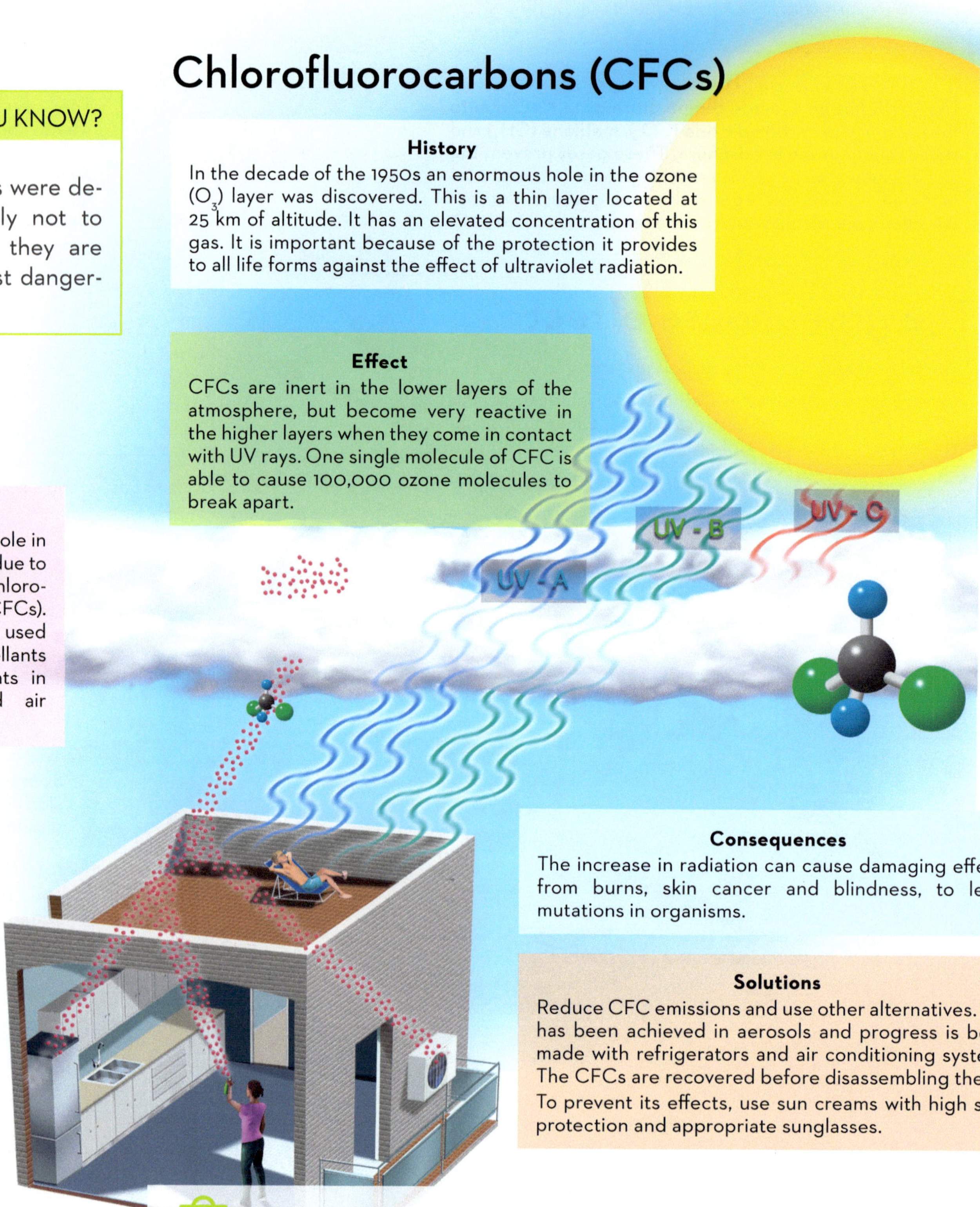

Consequences

The increase in radiation can cause damaging effects, from burns, skin cancer and blindness, to lethal mutations in organisms.

Solutions

Reduce CFC emissions and use other alternatives. This has been achieved in aerosols and progress is being made with refrigerators and air conditioning systems. The CFCs are recovered before disassembling them.

To prevent its effects, use sun creams with high solar protection and appropriate sunglasses.

ACTIVITIES

28. **Go to www.epa.gov/sunwise/es and access the publications. Look for sun protection measures.**

Then go to *children* and mark the icons on the right. Finally, do the activities indicated, from easy to difficult, and take a screenshot to show your achievement.

WORKSHEET

Download this unit's worksheet and do activities 25, 26 and 27.

8.1. Industrial waste

One of the most important effects of industrial culture is the appearance of **waste**. Excluding household waste, industry causes a great impact on the planet. Some of its effects are:

Environmental effect or problem		Solution
Plastics that never decompose can remain unchanged for centuries. There are floating islands of plastic materials in the oceans. These are traps for marine animals: turtles, dolphins, sperm whales, whales, etc.		Produce biodegradable or photodegradable plastics that are destroyed by microorganisms or sunlight. However, the best solution is to **reduce** consumption and **recycle** the material used.
Oil spills, fundamentally in the sea and on the coasts, affect flora and fauna. Animals and plants die by contact or by asphyxiation.		Improve fuel collection methods. The best option is not to transport it and encourage clean, **alternative energies**.
Accumulation of mining waste in tailing ponds that can break and release large amounts of very toxic substances. Nature cannot eliminate these toxins.		Minimise the extraction of minerals, preventing their excessive use through reducing, **reusing** and **recycling** devices.
Groundwater contamination due to the abuse of nitrogen-based fertilisers to improve agricultural production without adequate control.		Improve information for farmers to achieve an **appropriate use** of fertilisers, obtaining better-quality products.
Eutrophication of water is an increase in the nutrients in a body of water. This causes an increase in the amount of algae, plants and organisms. When they die, they pollute the water and make it uninhabitable.		**Decrease** the uncontrolled dumping of liquid contaminants, fertilisers, livestock excrements, forest residue from harvesting wood and unpurified urban waste.

ACTIVITIES

29. It may seem like chemistry is the cause of multiple types of pollution and that there are only two alternatives: completely abandoning it and going back to the ways of the past, or accepting the problems it causes.

    However, there is a third option that is much more advantageous: encouraging a rational, responsible use of chemistry, sources and products.

    Make a questionnaire with three questions on different aspects of each one of the environmental problems studied. Interview six people you know to find out about their level of knowledge on the subject.

    Analyse their answers, comparing similarities and differences according to age, educational level, gender, etc. Determine if they are aware of the problems posed. Using this information, prepare an oral presentation or report. Propose specific actions to combat each one of the problems studied and encourage positive attitudes in the people you know.

MIND MAP

Download and print the worksheet you will find in the OLC to complete the mind map of the unit.

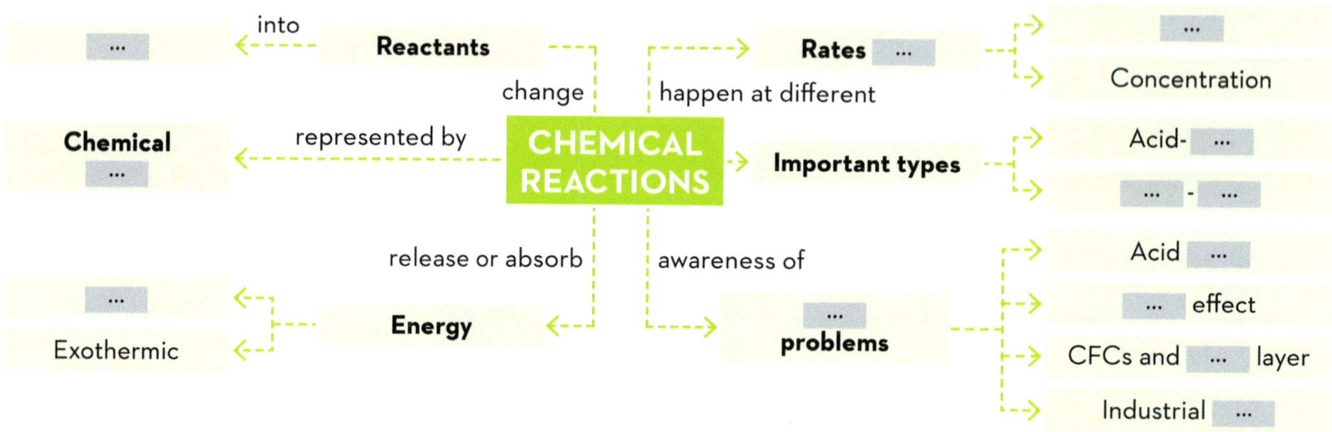

LOOK AROUND YOU

Five new materials of the possible immediate future

Materials science is making a lot of progress and its application is virtually immediate.

Absorb water... from the air: Researchers at Rice University in Houston have created a material capable of absorbing water from the humidity in the environment, however low it may be. By imitating [...], scientists have created a 'hygroscopic scaffolding', similar to a forest of carbon nanotubes capable of attracting water molecules from the air and capturing them inside. And, like a sponge, it can be drained and used again. When this material is ready to be produced in large quantities, it could enable the creation of fabric that retains moisture and permit the habitation of extremely arid areas, just like in the science fiction saga *Dune*.

Diamonds harder than diamond: Diamond is the hardest material, capable of scratching any other solid, which is why it is used in industry, mining and construction. Chinese scientists have published a paper in the prestigious journal *Nature* about the manufacture of an artificial diamond that is even harder, able to withstand pressures of nearly two million atmospheres, twice what would shatter a natural diamond. Also, it maintains its properties at high temperatures, which extends its applicability in industrial processes. [...] Researchers used carbon nanoparticles, matched and structured in layers like onions, as a nucleus for the diamond. They say that twice the hardness is achieved this way.

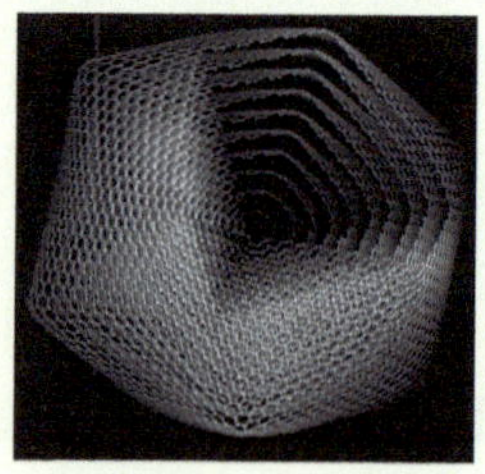

www.bbvaopenmind.com, 9.10.2014 (adaptation)

Questions

a) What does the material that attracts water imitate? What other new materials does the article talk about?

b) How is the material harder than a diamond made? Make a poster to compare them.

Reading techniques

Use the worksheet from the OLC to practise reading techniques that illustrate strategies to get information from a text.

LABORATORY WORK

Iodine stopwatch

Objective

You are used to reactions that show the desired effect instantly. Now you will do an experiment that is a chain of reactions. The final result is not immediate. As the time it takes to produce a change of colour depends on the amount of substance added, you can use it as a programmable stopwatch.

Material

- Sodium iodate ($NaIO_3$).
- Sodium hydrogen sulphite ($NaHSO_3$).
- Sulphuric acid (H_2SO_4) 1 M.
- Starch.
- Stopwatch.
- Beakers.
- Distilled water.

Procedure

You must prepare two solutions as indicated below:

- **Solution A:** Dissolve 0.5 g of sodium iodate in 50 mL of water. Use a pipette or a burette and divide this solution into three test tubes with 2 mL in each one of them. Label them with the codes A1, A2 and A3.
- **Solution B:** Add 1 g of starch and 10 drops of sulphuric acid to 50 mL of water and stir the mixture very carefully. Try not to splash the solution. Next, add 0.25 g of sodium hydrogen sulphite and stir again until it is completely dissolved. Divide this solution into three test tubes, but with the following amounts:

B1	2 mL of solution B	No water
B2	2 mL of solution B	2 mL of water
B3	2 mL of solution B	4 mL of water

- Start the stopwatch and quickly mix solutions **A1** and **B1.** Determine the time taken for the colour to change.
- Repeat the experiment with the other solutions that have the same number.

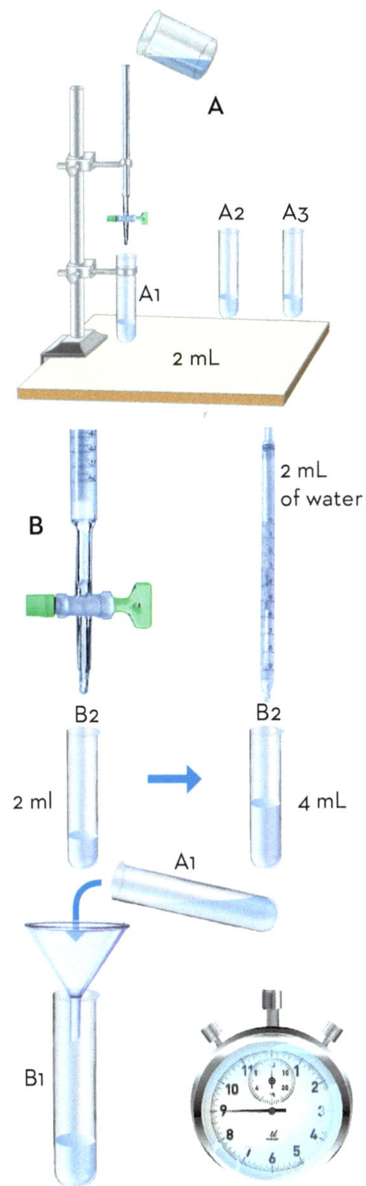

Task

Present the laboratory report and accompany it with a video demonstrating the results obtained. Search for possible applications and try to predict new times. If you have the opportunity, test your hypotheses.

Graph your results, showing the amount of water added in the horizontal axis and the time needed to complete the reaction in the vertical axis. From the shape of the chart, describe the relationship between concentration and reaction kinetics.

FINAL ACTIVITIES

Basic activities

1. Identify what type of transformation occurred in each situation: physical or chemical. Check with your partner.

a) Lighting a match.

b) Adding salt to water.

c) Turning on a fluorescent light bulb.

d) Ice cream melting with heat.

e) Melting tin.

f) An alkaline battery.

g) Making mayonnaise.

2. For the chemical transformations in activity 1, name the initial and final substances.

3. Describe what happens in the following reaction with the words: reactant, product.

$$H_2CO_3\,(aq) \rightarrow CO_2\,(g) + H_2O\,(l)$$

4. In the following reaction:

$$C_3H_8\,(g) + 5\,O_2\,(g) \rightarrow 3\,CO_2\,(g) + 4\,H_2O\,(g)$$

a) Identify the reactants and the products.

b) Indicate the broken and formed bonds.

5. In the following reaction:

$$C\,(s, \text{graphite}) + O_2\,(g) \rightarrow CO_2\,(g)$$

a) Identify the reactants and the products.

b) Indicate the broken and formed bonds.

6. In order to break a bond, must energy be provided or eliminated?

7. If the energy of the reactants is greater than that of the products, what happens in regard to energy? What is this called?

8. Potassium hydroxide is a substance that colours the indicator paper blue. What type is it? Would it react with orange juice?

9. What do we call the type of reaction that occurs inside a battery? Compare your answer with your partner.

10. We use vinegar to remove lime stains on taps and coffee pots. Explain why this works.

11. What mass of oxygen has been added to 222 g of iron if 318 g of iron (III) oxide is obtained? If we weigh the initial iron and the final oxide, will the Law of Conservation of Matter be fulfilled?

12. A chemical reaction needs to be accelerated. What is better?

a) Heating or cooling?

b) Concentrating or diluting the reactants? And the products?

c) Explain both answers according to the Corpuscular Theory of Matter.

d) Discuss your answers in small groups.

13. Working in pairs, create a table to classify the following substances according to their origin: natural, manufactured or synthetic:

a) Pen cap.

b) Toothpick.

c) Cotton sweater.

d) Red brick.

e) Car tyre.

f) Paper clip.

g) Nylon stocking.

h) Window glass.

i) Wine cork.

j) Paracetamol.

14. Identify ten synthetic products that have improved our quality of life. Share your list with the whole class, to create a big common list.

15. Look for three news articles about contributions of chemistry to our society. Make a presentation for your classmates about them and defend the benefits of using chemistry in a responsible society.

16. Describe the main environmental problems. For each one of them, propose three individual and three collective measures to minimise or eliminate their impact. Work in small groups. Use everyone's contribution to create a group environmental report. Use adequate academic language.

Consolidation activities

17. In the following reaction:

$$4\,Fe\,(s) + 3\,O_2\,(g) \rightarrow 2\,Fe_2O_3\,(s)$$

a) Identify the reactants and the products.

b) What type of bond does each one of the substances participating have?

c) Indicate the broken and formed bonds.

18. Why do certain medicines have to be stored in cool places?

19. You have probably observed that a greenish precipitate, copper chloride, appears on taps. Working in pairs, research why this substance appears.

20. The following map shows the main producers of nitrogen oxides in a curious way. What consequences come from them and what countries are the main producers? In a small group, debate about the meaning of the map and decide on the reasons for these shapes.

21. Two of your classmates have obtained 6.36 g of calcium hydroxide ($Ca(OH)_2$) from calcium oxide (CaO) and water, but they have written different values. Are they correct?

- Classmate A: 4.8 g CaO + 1.56 g H_2O
- Classmate B: 4.86 g CaO + 1.3 g H_2O

22. Take a photo of the labels on five articles of clothing in your wardrobe and classify them according to their origin. Share with the rest of the classroom to create a common classification with all the contributions.

23. A family member asks you to help him take a refrigerator to a dump. What environmental problem is he making worse? How would you dissuade him? What alternative would you suggest? Working in pairs, write a small theatre play and represent it in front of the whole group.

Advanced activities

24. Work in small groups and look for information to determine if the following transformations are physical or chemical. For the chemical transformations, determine the initial and final products.

a) Making cream from egg whites.

b) Applying super glue to an object.

c) Turning on a television set.

d) Washing clothes.

25. There are drinks sold in cups that heat up automatically when you shake them. What reaction occurs and what type is it?

26. Discover the acidic or basic nature of:

a) Strawberries. *b)* Rainwater.

c) Shampoo. *d)* Washing-up liquid.

27. There is a magnesium electrode inside electric heaters. Explain its chemical purpose. Debate with the whole group.

28. What would happen if all plastic suddenly disappeared? What alternatives are there? Debate with the whole group.

29. When you exercise energetically, your body does not have time to get enough oxygen. What mechanism does the organism use to obtain energy and what consequences does it produce?

30. Many additives have a natural origin and others are synthetic. Take a photo of the labels on five food products, identify their preservatives, dyes and other additives and find out if they are natural or artificial. Share with the rest of the classroom to create a group list with all the contributions.

31. Look for pictures of forests before and after acid rain. Make a poster with another student explaining what happened. Suggest actions to prevent pollution and solutions for when it is already present.

USE YOUR SKILLS

Bread

A baker makes bread by mixing flour, water, salt and yeast. When everything is mixed, he places the dough in a container for several hours for the process of fermentation to occur. During fermentation, a chemical change occurs in the dough: the yeast (a single-celled fungus) transforms the starch and sugar of the flour into carbon dioxide and alcohol.

Question 1

Fermentation makes the mix rise. Why does it rise?

a) It produces alcohol, which transforms into gas.

b) It produces a gas, carbon dioxide.

c) The single-celled fungi reproduce inside.

d) The fermentation transforms the liquid water into vapour.

Question 2

A few hours after making the dough, the baker weighs it and observes that its mass has decreased.

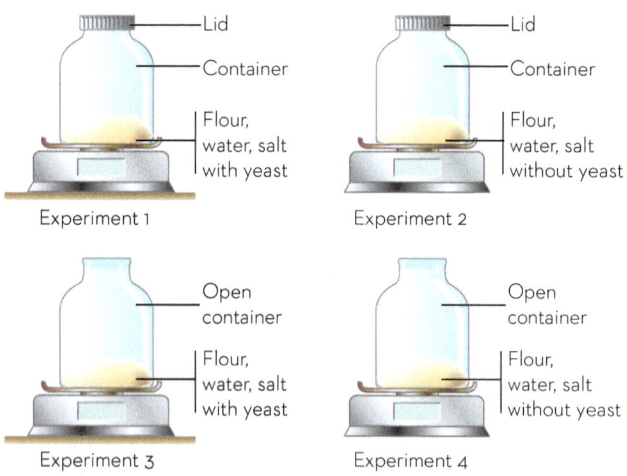

The mass of the dough is the same as at the beginning of each one of the four experiments shown. Which two experiments should the baker compare to determine if the yeast is responsible for the loss of mass?

a) Experiments 1 and 2.

b) Experiments 1 and 3.

c) Experiments 2 and 4.

d) Experiments 3 and 4.

Question 3

In the dough, the yeast transforms the starch and sugars of the flour into carbon dioxide and alcohol through a chemical reaction. Where do the carbon atoms that make up the carbon dioxide and alcohol come from? Answer *yes* or *no* for each one of the possible explanations.

a) Some carbon atoms come from the sugars.

b) Some carbon atoms were part of the salt molecules.

c) Some carbon atoms come from the water.

d) The carbon atoms formed from other elements in a chemical reaction.

Question 4

When the ripe (fermented) bread dough is baked in the oven, the bubbles of gas and vapour in the dough dilate. Why do the gases and vapours dilate when they are heated?

a) Their molecules get bigger.

b) Their molecules move faster.

c) The number of molecules increases.

d) Their molecules collide less frequently.

Check your answers to these questions with your partner. Explain your choice if it is different and listen to his/her explanation. Then, decide which one is correct.

The problem with plastics

→| Competency task

Objective

We are so used to them that we are not aware of their danger. Since their invention at the beginning of the 20th century, their use has increased exponentially. They have almost replaced natural materials in all areas of our lives.

But one of their advantages, stability over time, is making them one of the greatest environmental problems of the planet: **Plastic does not decompose** or it does so very slowly.

The consequences are terrible, especially for the marine environment where 'plastic islands' are beginning to be a serious problem for sea fauna and birds.

Final product

Although there are many possibilities, we suggest you make a video documentary or a digital presentation in public. Its mission is not only to exhibit the problem, but for the receivers of the message to reflect on it.

Step by step

Although your teacher may suggest a different process, we will indicate a possible sequence of steps. Working in groups and comparing the data obtained is recommendable.

1. Find information about the different types of plastic and their uses, especially those we use in our environment.
2. Identify the plastics you use in your house, what types they are, if they are biodegradable or not and the time they take to decompose.
3. Create a list of them at home and then meet up as a group to share.
4. Calculate the mass of plastic that the world produces.
   a) Determine the mass of plastic that the average family produces. Collect all the plastic you use in your house for a week, classify it according to its type and store it.
   b) Calculate the mass of each type of plastic that a person generates in a week. What will you have to do? Next extend it to one year and then to a whole lifetime (calculate 85 years).
   c) Make a bar graph. For each type of plastic, indicate the amount that one person has produced in one year. You can also add the time it takes for each type of plastic to disappear.
   d) Finally, determine the mass of plastic that the most developed societies produce (1.5 billion people).
5. Look for information on the plastic islands. What problem does this material cause for marine fauna, like birds?
6. Find information about the alternatives to current plastic use.

7. Do a series of interviews in which you show the interviewees the results of your study and collect their opinions.
8. Make the final product with all the information collected: data, calculations, visual aids, interviews, consequences and alternatives.

Self-assessment

Evaluate your work and your classmates' work with the guidelines that you can download from the OLC.

6 FORCES AND MOTION

'It is solved by walking.' You probably have heard this famous saying, but do we really know what motion is? Why do we move and how do we do it?

You belong to the first generation that considers space travel common. You are not surprised to hear that there are five probes on the surface of Mars or that there are astronauts circling the Earth every two hours.

One day not too far into the future, one of you will be on a cargo spaceship, nearing Europa, the moon of Jupiter with the highest probability of having liquid water. That day you will have to learn to stop a spaceship weighing over 2 000 tonnes that is heading toward the satellite at over 15 000 m/s, subject to a gravitational attraction that is increasing. And you have neither brakes nor a road!

Understanding how motion is produced and what magnitudes are used to analyse it will open up an unimaginable world for you, from the smallest motion, like the flight of a fly, to the largest motion, like the dance of the galaxies and the whole Universe. All of this has some very simple expressions that any thirteen-year-old person can comprehend.

'Research is to see what everybody else has seen and to think what nobody else has thought.'

Albert Szent-Györgyi, Hungarian-American biochemist.

Contents

WORK WITH THE TEXT

Working in pairs, answer the following questions about the text:

1. How do you say 'solvitur ambulando' in English?
2. What is Europa?
3. What makes Europa important?
4. Name at least four magnitudes that you think may be involved in the study of motion.

CHALLENGE

In order to conduct this experiment you need a container, a wooden spoon, a package of corn flour and a little water.

Put the package of corn flour into the container and add water gradually, stirring constantly. Continue until you have a thick mixture that is difficult to handle. Now, answer the following questions with the help of your partner:

- What state is the mixture in, solid or liquid?
- Now take a little in your hand and try to hold it. What happens? What state defines its behaviour?

- Now make a fist with your hand and hit the surface of the product with your knuckles. What happened? What state is it like now?

You can watch the following video to help you with your answers:

goo.gl/6Jxd5V

KEYWORDS

Download and print the activities that will guide you through the essential vocabulary of this unit.

FIND A CLASSMATE

Download and print the worksheet 'Find a classmate: road safety'. Share experiences about this topic with your classmates and discover interesting things about them.

1 Kinematics

VOCABULARY

The word kinematics comes from the Greek *kinema*, which means 'motion', and *-ics*, which indicates 'relating to'. So kinematics is the study of motion.

We are all used to the concept of motion, to the point that it is used as one of the characteristics that generally differentiates animals from other kingdoms like plants. The science that studies motion is called **kinematics**. But do you really know when you move?

THINK AND REASON

In the picture on the right, two friends are talking on the phone, one inside a car and the other inside a bus. Discuss with your partner:

a) Which of the two is moving?

b) Who is moving away from whom?

c) Can you find a fixed point?

In order to study the motion of a body, we need to define concepts:

Reference frame (RF): This consists of an oriented straight line in which motion occurs, called the **axis,** and a point that we accept is at rest, called the **origin** of the reference frame.

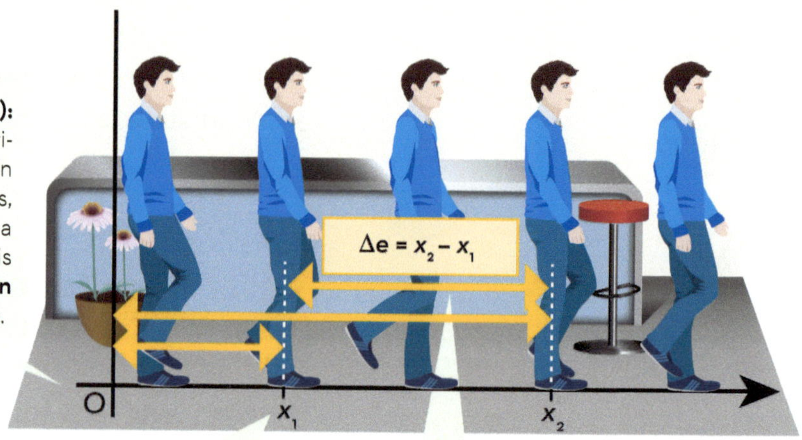

Position: distance from the object to the origin of the reference frame. Its sign depends on the side of the axis it is on.

Displacement: difference between any two positions.

A body **moves** when it varies its position in regard to the origin of the reference frame. It is called **moveable**.

ACTIVITIES

1. In the drawing in section 1, we will change the origin of the reference frame from the flower pot to the stool. Will the positions and the displacement change? Discuss your answers with your partner.

2. From the data in the figure, locate the position of the body at the starting time and at 2, 3 and 5 seconds.

a) How much has it moved between seconds 2 and 5?

b) How much has it moved in total?

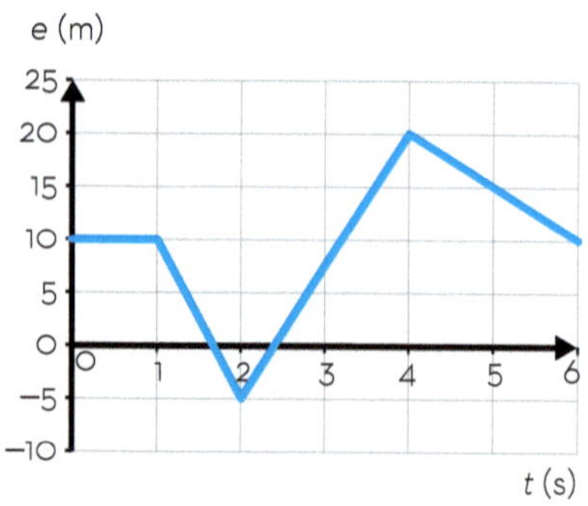

1.1. Velocity

Any two bodies can move on the same path but use different amounts of time. To differentiate this behaviour we use velocity.

Velocity is the measurement of the rate at which bodies move. It is calculated as the ratio between the displacement and the time used. Its units in the International System (SI) are m/s.

$$v = \frac{\Delta e}{\Delta t} = \frac{x_{final} - x_{initial}}{\Delta t}$$

REMEMBER

Although in everyday life it is common to measure velocity in km/h, the International System indicates that we should use metres and seconds in our calculations.

SOLVED EXAMPLE

1. Driving on a road at a constant velocity, you observe that you pass kilometre point 12 at 11:24 and kilometre point 21 at 11:30. What is the displacement? What is the velocity? When would you pass kilometre point 27?

Solution

We will make a drawing, locate the reference frame and identify data:

a) We determine displacement from the positions:

$$x_1 = 12\,000 \text{ m}$$
$$x_2 = 21\,000 \text{ m}$$
$$\Delta e = x_2 - x_1 = 9\,000 \text{ m}$$

b) Velocity is calculated by knowing the time used:

$$\Delta e = 9\,000 \text{ m}$$
$$\Delta t = 6 \cdot 60 = 360 \text{ s}$$
$$v = \frac{\Delta e}{\Delta t} = \frac{9\,000 \text{ m}}{360 \text{ s}} = 25 \frac{\text{m}}{\text{s}}$$

c) Using velocity, we can find the time we will take between $x_2 = 21$ km and $x_3 = 27$ km.

$\Delta e = 6\,000$ m
$v = 25$ m/s

$$v = \frac{\Delta e}{\Delta t}$$

Substitute variables and operate:

$$25 \text{ m/s} = \frac{6\,000 \text{ m}}{\Delta t}$$
$$\Delta t = \frac{6\,000 \text{ m}}{25 \text{ m/s}} = 240 \text{ s} = 4 \text{ minutes}$$

You will reach kilometre point 27 at 11:34.

WORKSHEET

Download this unit's worksheet and do activities 1, 2 and 3.

ACTIVITIES

3. In an 800-metre race, we observe that world champion David Rudisha passes the 150 m mark after 18 s and 75 hundredths.

a) How fast is he running?

b) What will be the final time of the race if he maintains this speed?

4. With the help of your partner, indicate which animal runs faster:

a) A wolf that runs 200 metres in 12 seconds.

b) A grey squirrel that runs 45 metres in 3.75 s.

1.2. Acceleration

Just as velocity measures the rate at which the position of a body changes, there is a magnitude that does the same with velocity. This magnitude is called acceleration.

Acceleration is the measurement of the rate at which the velocity of a body varies. It is calculated as the ratio between the variation of velocity and the time used. Its units in the International System (SI) are m/s^2.

$$a = \frac{\Delta v}{\Delta t} = \frac{v_{final} - v_{initial}}{\Delta t}$$

But what does an acceleration of 3 m/s^2 mean? It means that every second that passes, the velocity increases 3 m/s in regard to the velocity it already has. In short, the body moves faster and faster. If the acceleration is negative, velocity decreases in this value:

$a = +3$ m/s^2 (from 0 s to 3 s): +3 m/s, +3 m/s, +3 m/s

$a = -2$ m/s^2 (from 3 s to 5 s): −2 m/s, −2 m/s

Velocity	0 m/s	3 m/s	6 m/s	9 m/s	7 m/s	5 m/s
Time	0 s	1 s	2 s	3 s	4 s	5 s

SOLVED EXAMPLE

2. In order to take off, a commercial aeroplane must reach a velocity greater than 250 km/h, about 70 m/s. Determine the take-off acceleration if it reaches this velocity in one minute and 10 seconds.

Solution

We will make a drawing and identify the data:

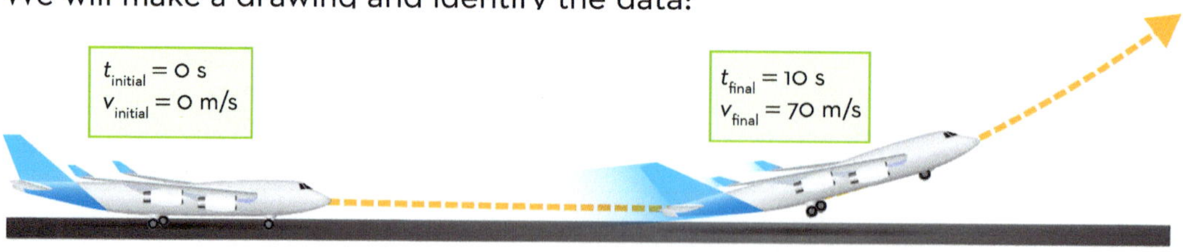

In order to determine the acceleration, we will use the expression:

$$a = \frac{\Delta v}{\Delta t} = \frac{v_{final} - v_{initial}}{\Delta t} = \frac{70 \text{ m/s} - 0 \text{ m/s}}{70 \text{ s}} = 1 \text{ m/s}^2$$

ACTIVITIES

5. A body moving at a velocity of 12 m/s has an acceleration of 4 m/s^2 for six seconds and then −6 m/s^2 until it stops. Working in pairs, make a table indicating its velocity at each second.

WORKSHEET

Download this unit's worksheet and do activities 4 and 5.

DID YOU KNOW?

The cheetah is the fastest animal and it also has the greatest ability to accelerate. It can go from 0 to 100 km/h in 3.5 seconds, whereas a sports car like the Mercedes-Benz SLR McLaren needs 3.8 s. Working with your partner, calculate the acceleration of the cheetah and the sports car.

1.3. Motion graphs

You know about the advantages of using graphic representation to visualise the variations of the magnitudes. This procedure is especially important in kinematics. Here are some examples:

A. Position-time graph

We can determine the **velocity** from them because they provide information on the position at each moment. Let's see an example:

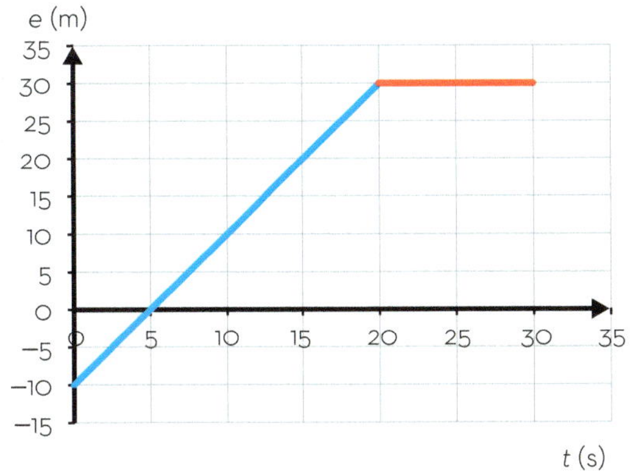

We can see two sections in the graph with different behaviours:

- **Blue** section: The position varies uniformly. It corresponds to motion with constant velocity. This is calculated from the initial and final points:

$$t_1 = 0 \text{ s} \rightarrow x_1 = -10 \text{ m}$$
$$t_2 = 20 \text{ s} \rightarrow x_2 = 30 \text{ m}$$
$$v = \frac{\Delta e}{\Delta t} = \frac{40 \text{ m}}{20 \text{ s}} = 2 \frac{\text{m}}{\text{s}}$$

- **Red** section: In this case, the position remains constant. This means the body does not move; it is at **rest**.

B. Velocity-time graph

Similarly to the previous graph, we can determine **acceleration**, which provides information on the velocity at each moment. Let's see an example:

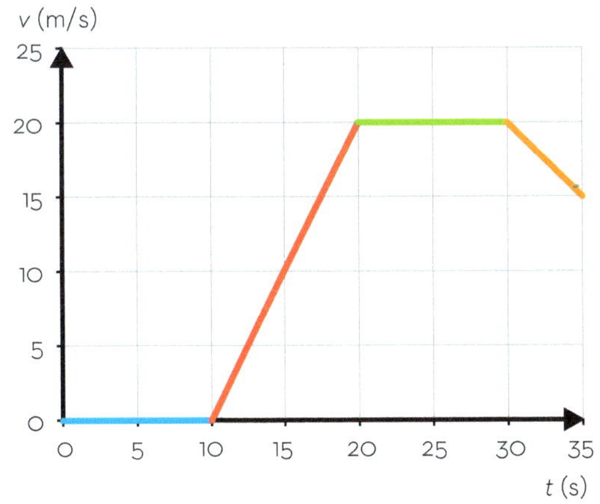

We can see four sections in the graph with different behaviours:

- **Blue** section: The velocity is zero and the body is at rest.
- **Red** section: The velocity varies uniformly. It corresponds to motion with constant acceleration. This is calculated from the initial and final points:

$$t_1 = 10 \text{ s} \rightarrow v = 0 \text{ m/s}$$
$$t_2 = 20 \text{ s} \rightarrow v = 20 \text{ m/s}$$
$$a = \frac{\Delta v}{\Delta t} = \frac{20 \text{ m/s}}{10 \text{ s}} = 2 \frac{\text{m}}{\text{s}^2}$$

- **Green** section: The velocity remains constant and the body continues moving at 20 m/s.
- **Orange** section: The velocity varies, decreasing from 20 m/s to 15 m/s in 5 seconds. Therefore, the acceleration is negative, -1 m/s^2, and the body decelerates.

PHYSICS 2.0

The following simulation will allow you to operate a 'moving man'. Observe the sign, position, velocity or acceleration he has depending on whether he is on one side or the other of the origin, whether he moves quickly or slowly, starts off or slows down.

goo.gl/qj6f4u

Another option is this maze game, where you can learn to go through it by using the magnitudes we have studied. Try to pass the test by manipulating the acceleration.

goo.gl/nu3sXW

WORKSHEET

Download this unit's worksheet and do activities 7 and 8.

1.4. Types of motion

In the previous sections you have seen that there are certain motions in which the velocity is constant: these are called **uniform motions**.

There are other motions in which the velocity varies at a constant rate. These are called **uniformly accelerated motions**.

SOLVED EXAMPLE

3. A car starts moving at a traffic light with an acceleration of 1.5 m/s² for eight seconds. Then it continues at a constant velocity for 20 seconds. At that moment it brakes with an acceleration of 2 m/s².

Solution

We will make a drawing and locate the reference frame at the traffic light. We will indicate all the data known in each section.

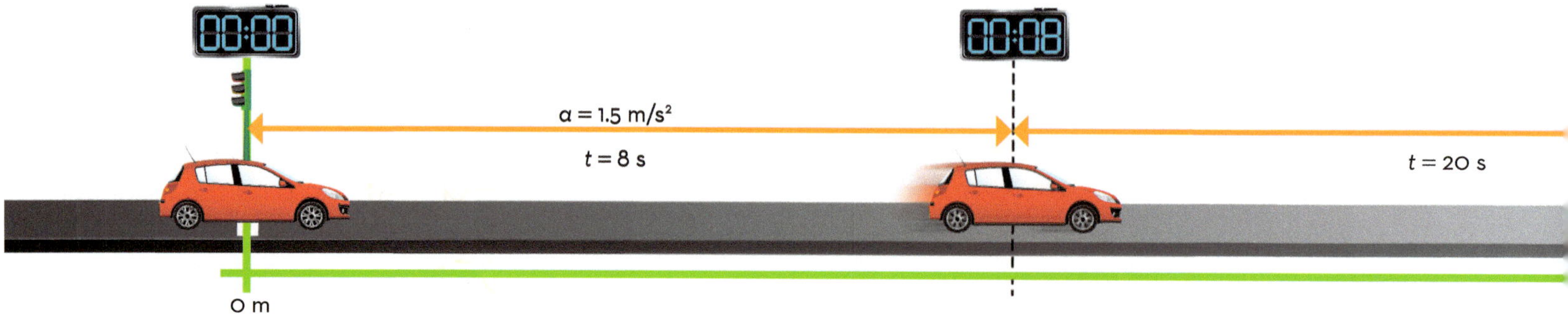

Uniformly accelerated motion

In this section there is a constant positive acceleration that causes the vehicle to increase its velocity from resting.

- Calculation of the variation of velocity:

$a = 1.5\ m/s^2$
$\Delta v = ?$
$\Delta t = 8\ s$

$$a = \frac{\Delta v}{\Delta t} \quad ; \quad 1.5 = \frac{\Delta v}{8} \quad ; \quad \Delta v = 12\ m/s$$

Because it starts from resting, its final velocity will be 12 m/s.

Uniform motion

The velocity is maintained, so its acceleration is zero.

- Calculation of the position:

$v = 12\ m/s^2$
$\Delta e = ?$
$\Delta t = 20\ s$

$$v = \frac{\Delta e}{\Delta t}$$

It has moved 240 m in the section.

CLASSROOM LABORATORY Uniform motion

1. Working in pairs, make a device like you see in the figure. From the origin of the reference frame make marks every twenty centimetres, as far as you can. Next prepare the reference frame from the highest position possible of the weight holder and let it drop.

   Determine the time it takes to move from the origin of the reference frame to the first mark. Take this measurement three times and find the average.

   Repeat the same process with all the marks, make a position-time graph and calculate the velocity of all the sections.

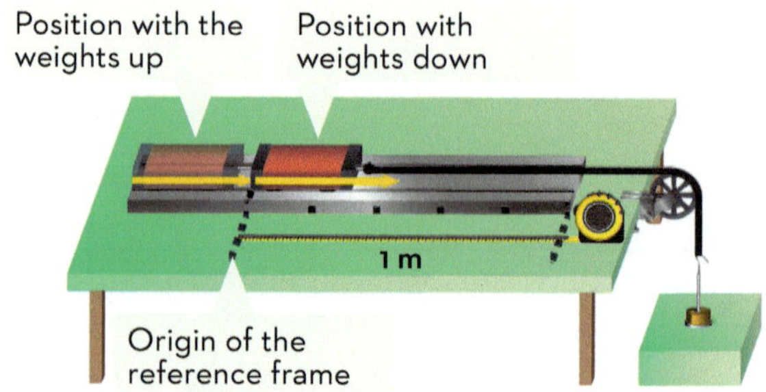

- $v = 0$ m/s — Rest
- $v \neq 0$ m/s
  - Constant velocity — $a = 0$ m/s² — Uniform motion
  - Non-constant velocity — $a \neq 0$ m/s² — Uniformly accelerated motion

Determine: a) The velocity of the first section. b) The distance moved in the second section.
c) The time it takes to stop.

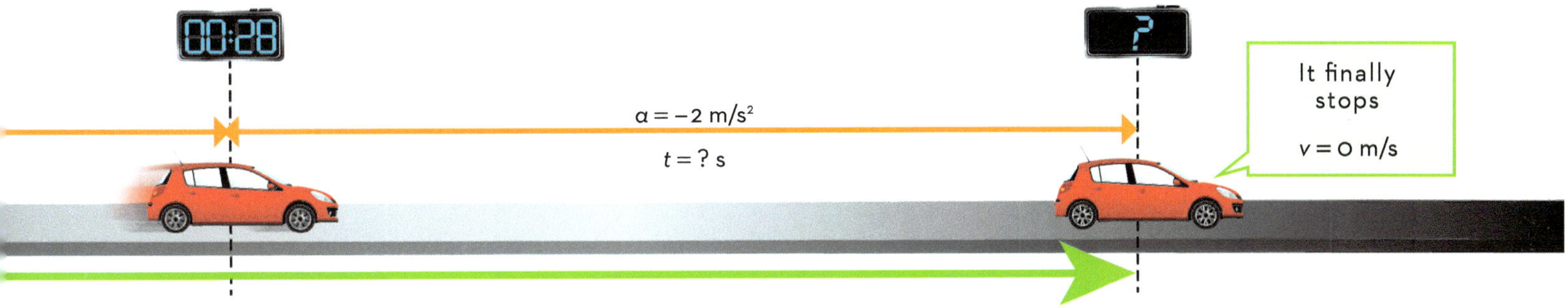

$12 = \frac{\Delta e}{20}$;

$\Delta e = 240$ m

Uniformly accelerated motion

In this section there is a constant **negative** acceleration that will cause the vehicle to decrease its velocity until rest:

- Calculation of the time. The velocity varies: $v = v_2 - v_1 = 0 - 12 = -12$ m/s

$a = -2$ m/s²
$\Delta v = -12$ m/s
$\Delta t = ?$

$$a = \frac{\Delta v}{\Delta t} \;;\; -2 = \frac{-12}{\Delta t} \;;\; -2 \cdot \Delta t = 12 \;;\; \Delta t = \frac{-12}{-2} = 6 \text{ s}$$

It takes 6 s to stop, and from the start it takes a total of 34 s.

ACTIVITIES

6. When a body falls to Earth, it accelerates 9.8 m/s². Working with your partner, determine the velocities it acquires every second for the falling times of the first five seconds.

Convert the velocities into kilometres per hour and make a velocity-time graph.

Match the falling velocities with the damage caused in a traffic accident according to the velocity. Use phrases like 'crashing at 20 km/h is equivalent to falling from a height of…'.

Make a large poster and show your results, explaining how you got them. Try to motivate people to drive cautiously.

2 The concept of force

We have studied how a body moves, but why does it move, stop or deform? What is the cause of these changes?

THINK AND REASON

Imagine you are leaning on a wall, as if you were pushing it. What happens to your hands?

Now imagine that the wall disappears. What happens to you? Has the action you are doing changed?

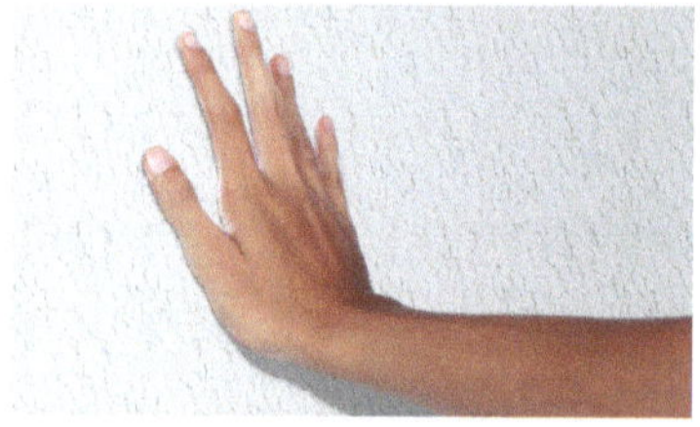

WORKSHEET

Download this unit's worksheet and do activities 9 and 10.

We all have a clear notion of force, but we cannot see it, only observe how it acts and the effects it produces. Let's see some examples:

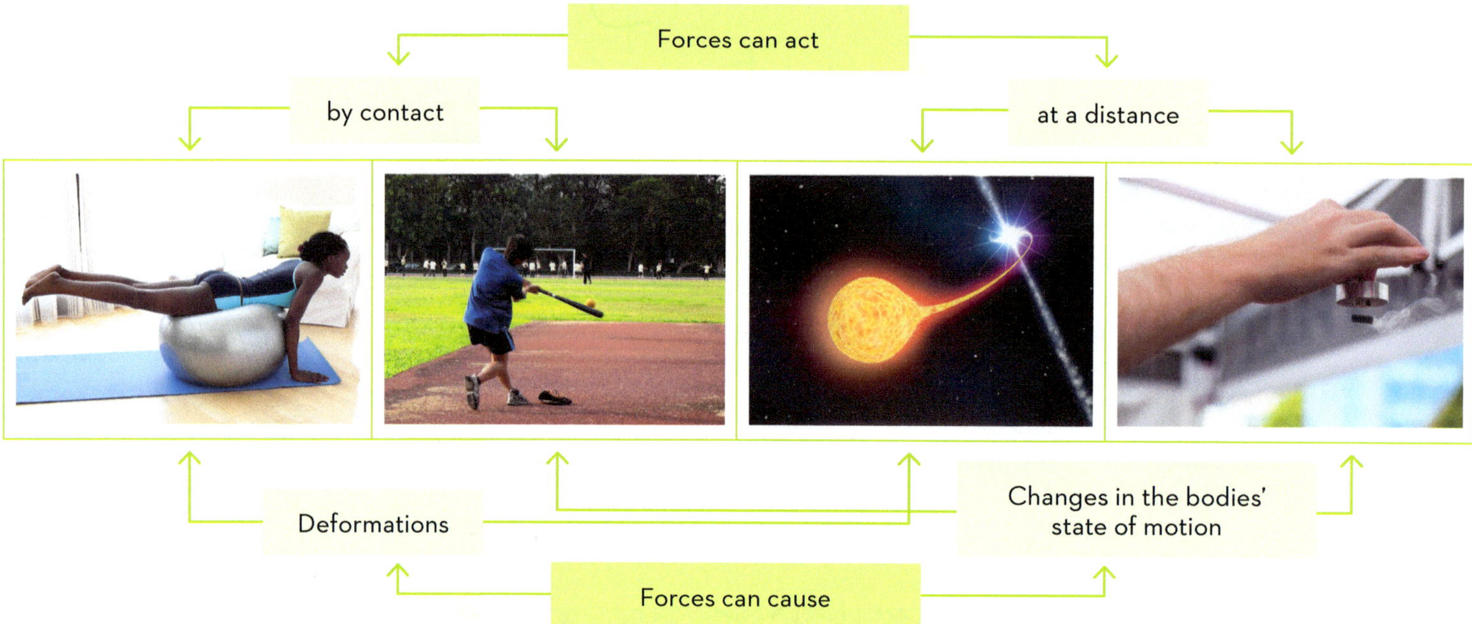

We define **force** as any interaction between two bodies by contact or at a distance that causes a deformation or a change in their state of resting or motion. Its unit of measurement is the newton (N).

DID YOU KNOW?

A newton is approximately the weight of a 102 gram body on the surface of the Earth.

ACTIVITIES

7. In the following situations, identify whether the forces have acted by contact or at a distance and what the effect was:

a) Kicking a ball.

b) Attracting a paper clip with a magnet.

c) A magnetic levitation train moving.

d) Squeezing grapes to make wine.

2.1. Forces have direction

If you take a chair and apply the same force in different places, is the effect achieved always the same?

As you have been able to see, the result of a force depends on many factors. In order to show this graphically, we will use a segment with an arrowhead, called a **vector**:

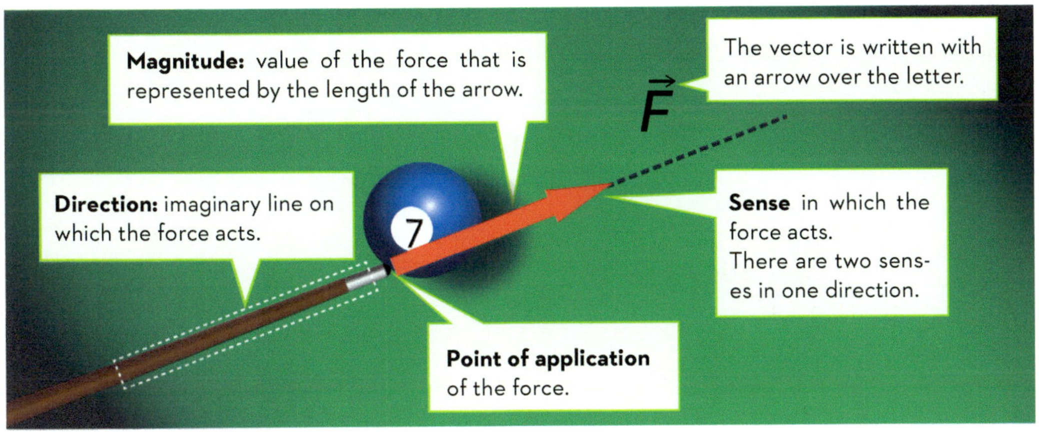

Composition of forces

A body can be applied various forces simultaneously. In these cases, it is possible to determine what is known as the resultant force.

> When various forces $\vec{F}_1$, $\vec{F}_2$, $\vec{F}_3$, etc. act on a body, it is possible to determine a **resultant force** $\vec{F}_T$. Its effects are the same as the combined action of all the individual forces.

This year we will only calculate resultant forces offorces whose vectors are on the same line, called the **line of action**. There are two situations:

Forces in the same direction

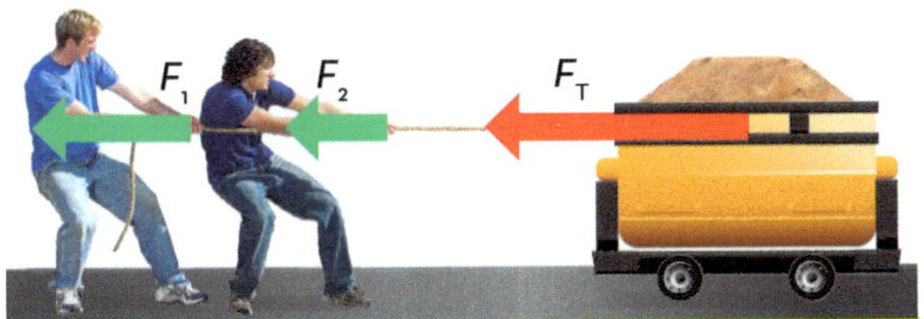

The forces are added together. Same sense.	$F_T = F_1 + F_2$

Forces in the opposite direction

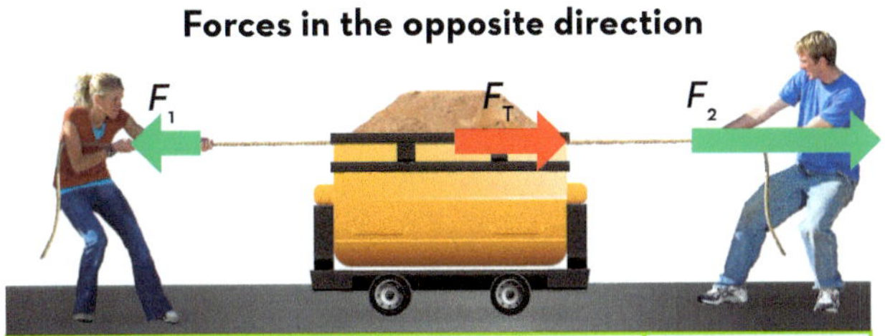

The forces are subtracted. Sense of the greater force.	$F_T = F_2 - F_1$

PHYSICS 2.0

You can practise more on composition of forces in the following PhET applet:

goo.gl/FgRgEr

ACTIVITIES

8. Draw the line of action of the forces and determine the total force exerted by the girls and boys in the following pictures:

200 N 250 N

100 N 40 N 60 N 60 N

PHYSICS 2.0

Many times nature imposes hard conditions on the constructions of human beings. Under these circumstances, the best option is to deform and adapt to them.

In the following link you will find a collection of curious situations where deformation prevents greater disasters.

goo.gl/4LLkg3

2.2. Deformations

A cushion is flattened when you sit on it and a bungee cord is stretched when you jump. In both cases the force causes a deformation.

A body has been **deformed** when a force is applied and its length varies along the force's line of action.

In order to analyse the causes, we will observe matter at the atomic level:

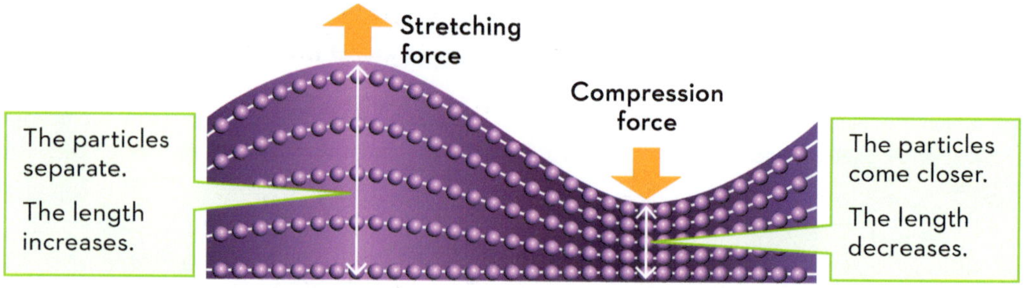

There are three types of bodies according to the behaviour they demonstrate with respect to deformation:

RIGID BODIES

They do not deform under the action of a force. As a consequence of this behaviour, they **break** or **fracture.**

PLASTIC BODIES

They deform when a force is applied to them, but they **do not regain** their initial length when the force disappears.

ELASTIC BODIES

They deform when a force is applied to them, and they do regain their initial length when the force disappears.

ACTIVITIES

9. Interpret how the particles of rigid, plastic and elastic bodies behave when a force is applied to them. In your notebook, draw what they look like before and after applying the force.

10. Identify the type of behaviour demonstrated by the objects in the pictures.

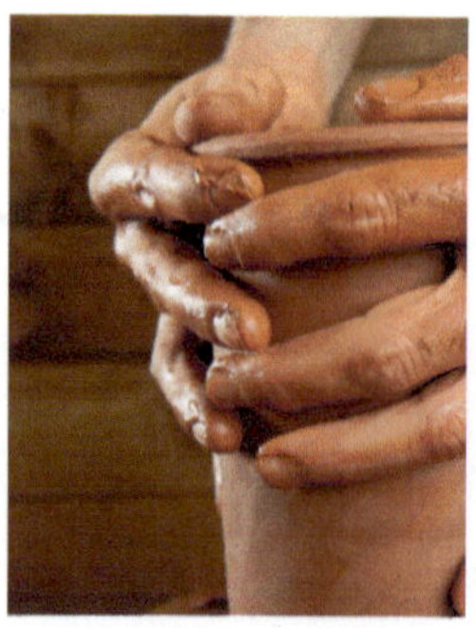

2.3. Hooke's law

You have probably observed the behaviour of a hand catapult, in which the bands stretch when you apply force to them. Do you think there is a relationship between the force applied and how much the hand catapult stretches?

WORKSH

Download sheet and do activities ..., 12 and 13.

CLASSROOM LABORATORY **Elongation of a spring**

2. Take a spring or another elastic object, hang it from a stand and put a weight holder on the other side.
   a) With a tape measure, determine the total length of the spring with the weight holder. We will call its value the **initial length** (ℓ_o).
   b) Add a **mass** weight and determine the **length** (ℓ) again.
   c) Continue adding weights until you have six measurements in total.
   d) Complete the following table, where $\Delta\ell = \ell - \ell_o$ corresponds to what is called **elongation** ($\Delta\ell$) and measure how much the spring has lengthened. This is calculated by subtracting the length with each weight from the length at the beginning.
   e) The last column corresponds to the ratio between mass and elongation.

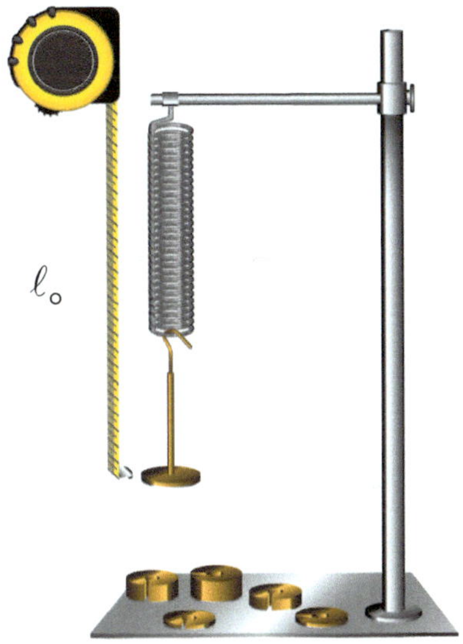

Mass (g)	ℓ (cm)	$\Delta\ell$ (cm)	$\frac{mass}{\Delta\ell}$
0	ℓ_o	0	---

What happens with the ratio between mass and elongation in all the cases?

As we will see later on, the weight of a body is a force that is proportional to the mass it has. Robert Hooke used this relationship in 1660 to state his law:

> **Hooke's Law** indicates that the force applied to an elastic body is directly proportional to the deformation caused. The ratio between the **force** (mass) applied and the **elongation** ($\Delta\ell$) is constant and it is called the **spring constant**.
>
> $$k = \frac{F}{\Delta\ell}$$

We can use this behaviour and build an instrument that allows us to measure the force: the **dynamometer.**

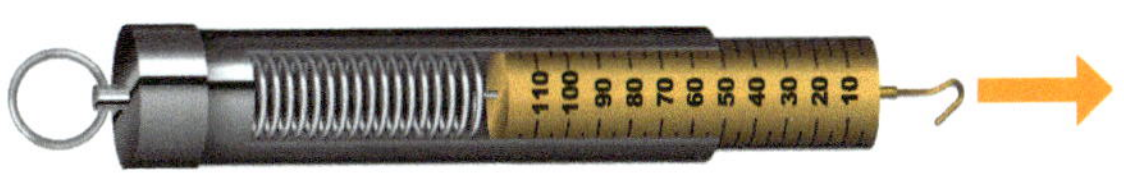

- The spring stretches by applying a force to it.
- The scale directly indicates the force applied in newtons (N).

PHYSICS 2.0

If you do not have springs or bands to do the activity, you can use the following simulator:

goo.gl/Lw2cJN

Try to do the activity with the three springs.

Remember that you can move the ruler and position it on the dotted line. This way you can directly measure the elongation.

ACTIVITIES

11. **Determine the spring constant of a 15 cm spring that measures 25 cm when a force of 200 N is applied. What will be its length when 350 N is applied? What force will stretch it 15 cm?**

2.4. Variation of the state of motion

When we push a chair or pull on a rope attached to it, it moves. In the first sections of the unit we studied how a body moves. Now you will learn why. The first question that you will answer is this: What does a force produce: velocity or acceleration?

THINK AND REASON

You are stopped on your bicycle. If you start to pedal by exerting force on the pedals, what happens to the bicycle?

If you stop pedalling, is your velocity zero?

Isaac Newton answered this question in 1687 by stating:

When a force $\boldsymbol{F}$ is applied to a body of mass $\boldsymbol{m}$, this acquires an acceleration $\boldsymbol{a}$ and the following expression is fulfilled: $\boldsymbol{F = m \cdot a}$

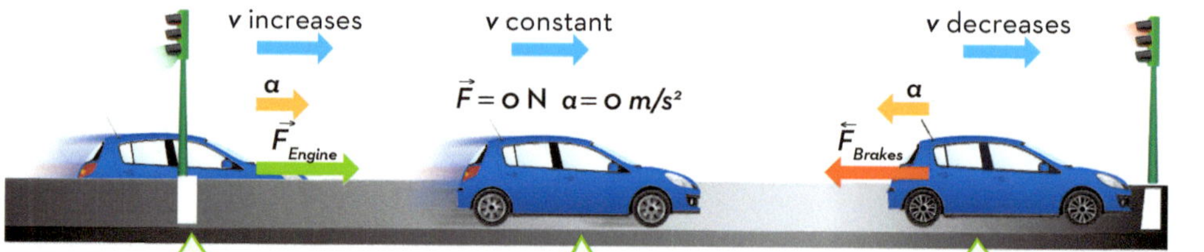

The force works with motion, it accelerates and velocity increases.

There is no force and velocity is maintained.

The force works against motion, it decelerates and velocity decreases.

SOLVED EXAMPLE

4. Two friends and you manage to push a car that acquires an acceleration of 1.25 m/s². What mass does the car have if you each exert 500 N of force?

Solution

As you all exert force in the same direction and sense: $F_T = 1\,500$ N

Data	Formula	Substitution	Solution
$F_T = 1\,500$ N $a = 1.25$ m/s²	$F_T = m \cdot a$	$1\,500 = m \cdot 1.25$	$m = \dfrac{1\,500}{1.25} = 1\,200$ kg

Consequently, the car has a mass of 1 200 kg.

! IMPORTANT

The procedure for solving a physics exercise is:

1. Read the exercise.
2. Extract the data.
3. Set out the formula.
4. Substitute the known values.
5. Solve for the unknown variable.
6. Answer what is asked.

PHYSICS 2.0

You can practise the relationship between force and motion in the following simulator. Activate the different graphs and push the filing cabinet.

goo.gl/QTeUVW

As it is probable that in the not-too-distant future you will pilot a spaceship, you should start practising.

Would you be able to land a spaceship with all its crew?

Go to the following link:

goo.gl/kiUgMp

ACTIVITIES

12. If we push with a force of 300 N on a body with a mass of 3 kg, what acceleration does it acquire?

13. Working in pairs, determine the force needed to make a car with a mass of 1 000 kg acquire an acceleration of 3 m/s².

3 Friction

When is it easier to slip, on wet surfaces or on dry surfaces? What about with trainers or shoes with smooth soles? Friction is a force that is present in our lives. You can make a clear conclusion from experience.

> **Friction** is a force that **always** opposes motion.

There are two types of friction:

- Fluid friction: This is what happens when a body moves in a fluid like air or water. It is due to the opposition of the fluid particles to separating. Its calculation is complex. Generally, it increases with velocity, which is why it gets harder for a car to accelerate the faster it goes.

- Solid Friction: This is what happens when two solid bodies come into contact. This is why it is difficult to drag objects. It is due to the irregularities of the bodies that come into contact: as they stick together, their relative motion is restricted.

  If water or oil is added between the bodies, this fills in the gaps, creating a cushion which decreases friction.

Friction plays a double role for us depending on the situation:

Advantageous

Tyres must grip the road in curves so that they do not slip.

There is friction.

Pushing furniture is more difficult on rough surfaces.

Magnetic levitation trains can travel at 400 km/h.

The friction is low.

Walking or driving a car on water or ice is very difficult.

Disadvantageous

PHYSICS 2.0

You can observe how the interaction that causes friction occurs in the following simulation.

goo.gl/kPCPds

WORKSHEET

Download this unit's worksheet and do activities 14, 15 and 16.

ACTIVITIES

14. Why do regular car tyres have a tread whereas race car tyres are bald? Discuss your answer with your partner.

15. If we push a 150 kg filing cabinet with a force of 400 N, what acceleration does it acquire if friction is 150 N? What about if there is no friction?

16. With the help of your partner, identify five situations where the presence of friction is advantageous and five where it is disadvantageous. Take photos or draw pictures and make a poster, explaining all the situations in an organised way. Then explain your poster to the class.

4 The force of gravity

Probably the first force that we learnt about was that if you stumble, you fall. The Earth attracts all the bodies that are in its radius of action, but this attraction is not limited to our planet.

THINK AND REASON

From what you have seen on television and in films, imagine the answer:

a) It is harder to pick up a whole log than if we cut it in half. What would we feel if half of the Earth suddenly disappeared?

b) As we go up and away from the Earth, will it be harder or easier to move away from the Earth? Can we turn off the engines of the spaceship we are travelling in?

You probably know about Isaac Newton and the story of the apple (which did not fall on his head). From the observation of the fruit falling and the movement of the Moon around the Earth, the scientist came to the following conclusion:

Gravity is a force that acts at a distance between two bodies that have mass. It increases with the value of these masses and decreases with the distance that separates these bodies.

This force has four important characteristics:

Measuring the force of gravity is easy. It can be done with a dynamometer or with a scale. Both instruments have a spring inside that deforms from the action of the body's weight force. The difference between the scale and the dynamometer is that the dynamometer usually gives readings in newtons and the scale is calibrated in kilograms.

ACTIVITIES

17. **Discuss with your partner. If we get on an aeroplane and stand on a scale for the entire flight, what would happen throughout the flight from take-off to landing?**

PHYSICS 2.0

Here is a collection of curious videos that show you what life would be like without gravity.

Try to find out what motions are produced:

goo.gl/fwPcBW

WORKSHEET

Download this unit's worksheet and do activities 17 and 18.

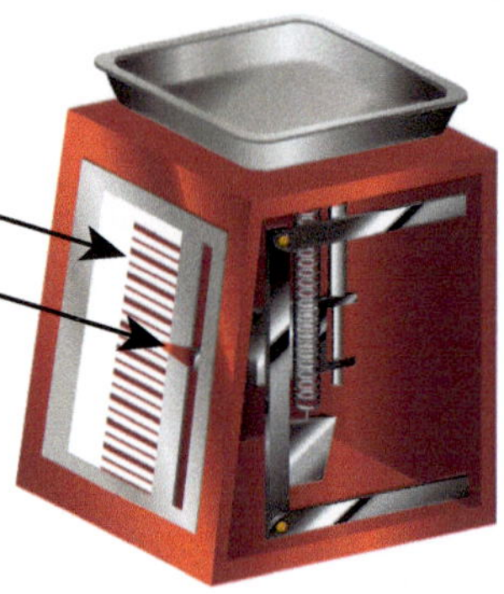

4.1. Weight

When we are near the surface of a planet, gravity has a special name: **weight force**. Its calculation is simple:

$$W = m \cdot g$$

where m is the mass of the body in kilograms and g is known as gravitational acceleration. Its value depends on the planet where the body is located. On Earth, gravitational acceleration, or simply, gravity, is 9.8 m/s^2.

Like all forces, there can be two effects:

Deform

If the body that experiences the force is a solid, it usually flattens and may break.

If the system is a fluid, it will change its shape, like the tides caused by the attraction of the Sun and the Moon.

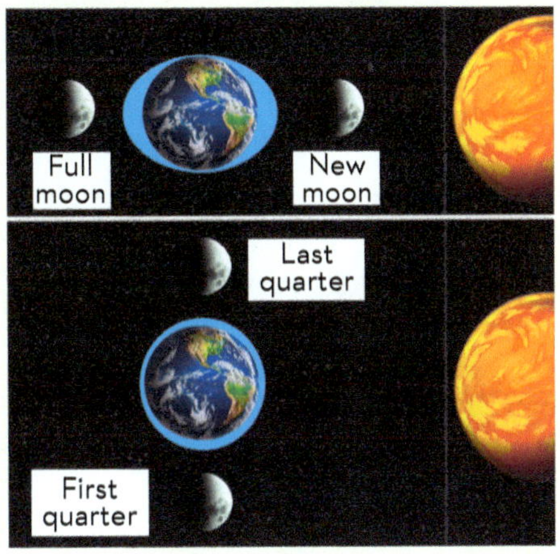

Vary the state of motion

If the body is free, this acceleration will cause its ascent to slow down and/or initiate its descent.

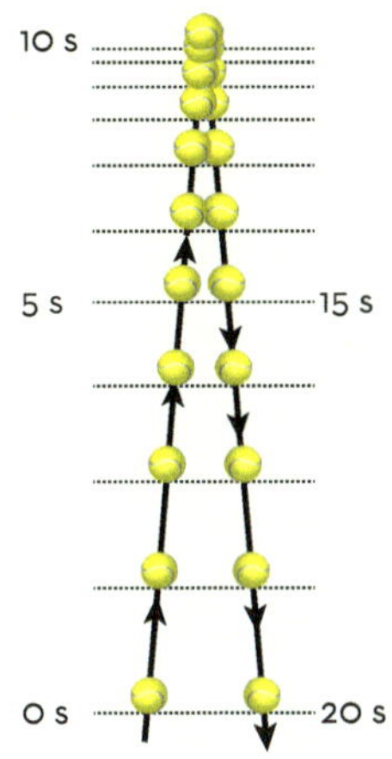

EXPERIMENT The force of gravity

1. We explained that the force of gravity increases with the mass of the body. So, what will reach the floor first, a sheet of paper or a rubber?

   With the help of your partner, drop a sheet of paper and a rubber while standing on a chair. With a camera (the ones in mobile phones are too slow), try to photograph the place of impact. Now repeat this, but crumple up the paper in the shape of a ball.

   What happened with the gravitational acceleration that they experience? Make a poster and use your photographs to explain what happened. Present your conclusions to the rest of the class.

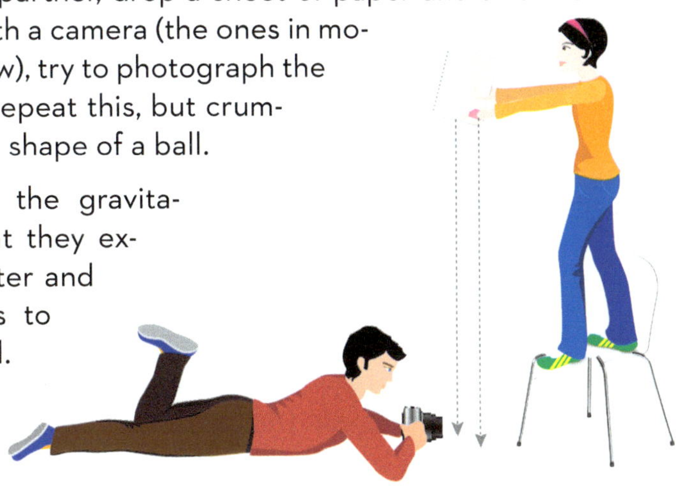

PHYSICS 2.0

Human beings have confirmed experiment 1 by going to the Moon, and it is not a lie. Observe how an astronaut does the same experiment in:

goo.gl/MqxzEu

4.2. Mass, weight, gravity and universe

Note that mass is not the same thing as weight. Mass is a measurement of the amount of matter that a body has. It is measured in kilograms and it does not depend on the place where it is measured.

Weight is the force with which a planet attracts us. It is measured in newtons and its value depends on where you are.

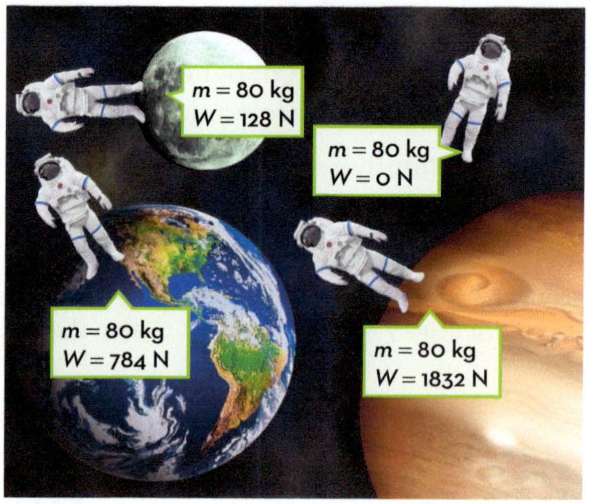

If we are far from any body that has mass, we will not feel the force of attraction. We will be **weightless**.

The force of gravitational attraction also acts between planets and stars and is responsible for the shape of the Universe: from the formation of the stars and planets to the constitution of the solar systems and the galaxies.

If the bodies rotate fast enough, the force of gravity will prevent them from separating and they will continue rotating forever.

DID YOU KNOW?

Solve the equation

The gravity of each planet depends on its mass. Look at their values.

Mercury	2.8 m/s²
Earth	9.8 m/s²
Moon	1.6 m/s²
Mars	3.7 m/s²
Jupiter	22.9 m/s²
Sun	274 m/s²

PHYSICS 2.0

You can see a simulation of the planets' orbits in the following link:

goo.gl/Qs8nLg

Activate: *show gravity force and path* and emulate the Sun/Earth/Moon system.

Try eliminating gravity for less than a second. Does the Earth come back?

ACTIVITIES

18. Calculate your mass and your weight on the Earth, Moon, Mars, Jupiter and the Sun. Compare your results with your partner.

19. Our necks can hold up between 80 and 100 kg with practice. If our heads weigh around five kilograms, what is the maximum gravity that our necks can tolerate? (Remember that force is measured in newtons.)

20. Determine the gravity of a planet, knowing that 5.3 kg weigh 23.32 N. If the planet is rocky, will it have more mass than the Earth? Discuss your opinion with your partner.

4.3. Distances in the universe

In the past, human beings measured large distances according to the time it took to travel them: a day's journey, a two-week voyage, etc. With the appearance of new means of transport, it was necessary to measure them in kilometres in order to avoid mentioning the vehicle.

However, distances in space are so large that working with kilometres is unviable. Like in the past, it is easier to use the travel time as a reference. On this occasion, the traveller will be light, which travels at the incredible velocity of 300 000 km/s.

So we have a new measurement of **distance:**

Light-time is the distance that light travels in that time.

SOLVED EXAMPLES

5. The Moon is 384 400 km from the Earth. How far away is it in light-time?

Solution

Knowing the distance and the velocity, we only need to apply kinematics:

$$\left.\begin{array}{l} e = 384\,400 \text{ km} \\ v = 300\,000 \text{ km/s} \end{array}\right\} \quad e = v \cdot t \qquad t = \frac{e}{v} \qquad t = \frac{384\,400 \text{ km}}{300\,000 \text{ km/s}} = 1.28 \text{ s}$$

The Moon is 1.28 light-seconds away, that is, its light takes 1.28 seconds to reach us. This is so far away that all the planets of the Solar System would fit inside this distance (**goo.gl/VWhW4G**).

From light-time we have the following unit of distance:

A **light-year** is defined as the distance that light travels in a year. It is equal to 9 461 000 000 000 km.

ACTIVITIES

21. The Sun is 149 600 000 km away. How long does it take its light to reach the Earth? If the Sun disappeared at 12:00 am, what would you observe for the following ten minutes? Discuss your answer with your partner.

22. How many kilometres separate us from our nearest star, Alpha Centauri, if it is 4.3 light years away? Currently, the fastest spaceship travels at 17 km/s. How long would it take to get to that star?

23. The nearest galaxy to ours is Andromeda. Working in pairs, answer the following questions: How far away is it? *Did Homo sapiens* exist when the light that reaches us now left there?

PHYSICS 2.0

The following simulation will help you visualise how far away Mars is based on the size of your screen.

www.distancetomars.com

The following simulation will help you move throughout the solar system, using the mouse wheel or the space bar.

goo.gl/K1RJLM

Isn't it difficult to reach Saturn?

WORKSHEET

Download this unit's worksheet and do activities 19 and 20.

PHYSICS 2.0

A great way to visualise the size of our Solar System, the planets in it and the Universe is in the following page of videos:

goo.gl/Lnf8Kz

Was it how you imagined?

5 Electric force

REMEMBER

Atoms are neutral; they have the same number of protons (positive charge) as electrons (negative charge).

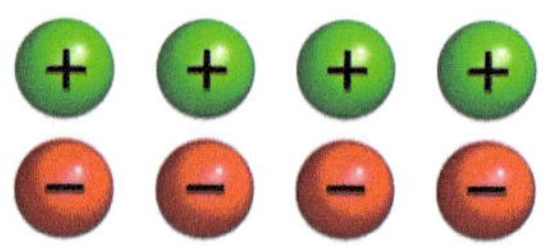

When an atom gains electrons it acquires a **negative** charge.

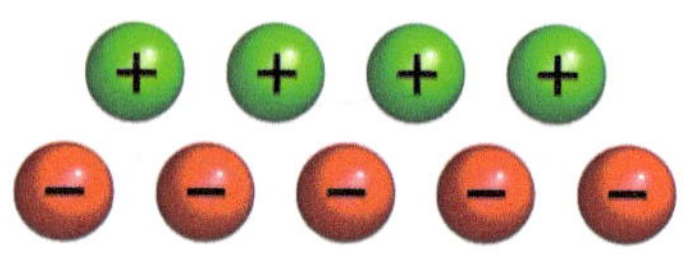

When an atom loses electrons it acquires a **positive** charge.

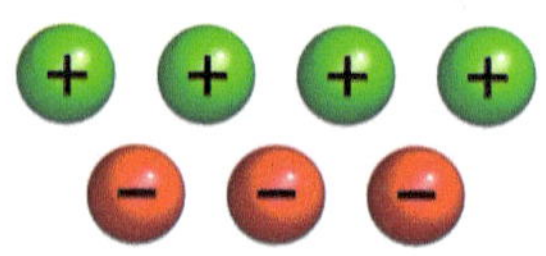

You have studied that matter is made up of atoms formed by a nucleus surrounded by electrons. When these electrons are gained or lost, for example, by rubbing, the matter acquires an electric charge. But how does it behave?

CLASSROOM LABORATORY Together or separate?

3. Working in pairs, collect the material indicated. Blow up the balloons, tie each of them to one end of the string and put the string over a ruler.
   - What happens to the balloons?
   - Rub just one of the balloons on a wool jumper and let it go. What happens now?

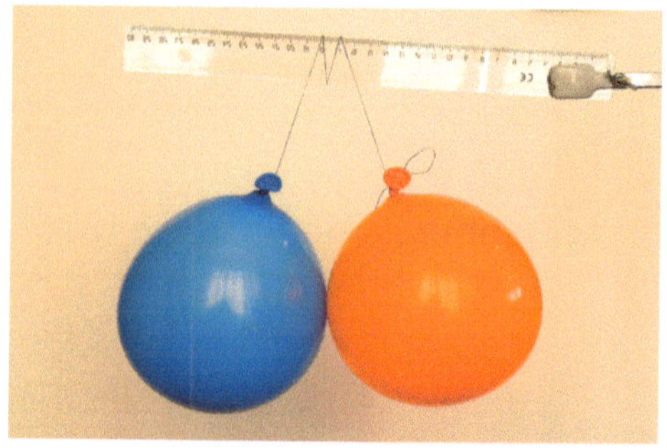

- Rub both balloons with the same wool jumper and let them go. What happens?

When two bodies are charged, a force appears between them with the following characteristics:

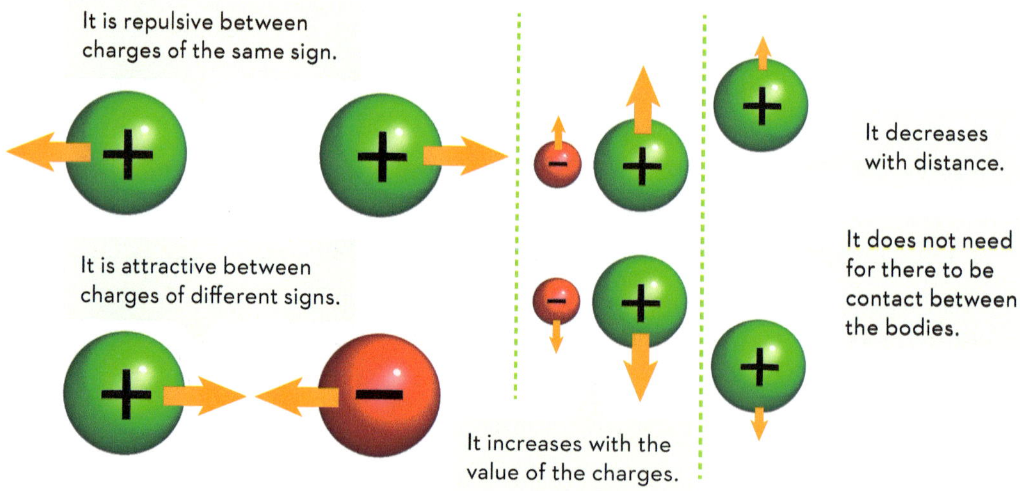

The charges of a body have a certain capacity for motion and two situations may occur:

All the charges that have left at the moment are linked to a body. They cannot leave it and if they do, they will redistribute until their motion ceases. This is what is known as **static electricity**.

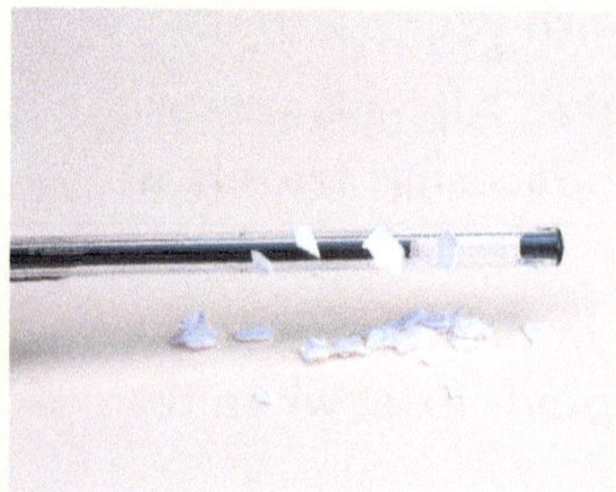

If we add an element that impulses the charges continuously, they will move constantly. This flow of charges is called **electric current**.

5.1. Similarities and differences with gravity

Electric force and gravitational force are similar, but not the same:

They are similar	They are different	
	Gravitational force	**Electric force**
They act at a distance.	It is always attractive.	It can be attractive or repulsive.
Their value decreases when the bodies are separated.	The masses are always positive.	The charges can be positive or negative.

WORKSHEET

Download this unit's worksheet and do activity 21.

5.2. Static electricity in our life

The manifestations of static electricity are numerous. It is easy to identify them and we can classify them according to how the electric charge is acquired:

By rubbing	By induction	By contact
This happens when two bodies in contact slide together. The charges of one go to the other.	When a charged body-comes near a neutral body, it induces a separation of charges.	When a charged body touches a neutral body, it gives the neutral body part of its charge.
Combing dry hair.	Hair nearing a television.	Touching a car.

EXPERIMENT The flying ring

2. We need a plastic bag and a blown-up balloon. Pull the bag out and fold it lengthwise in four parts. Then make two cuts five and eight centimetres from the ends and unfold it to make a ring.

Rub the balloon and the ring against your hair or a wool jumper, throw the ring in the air and try to make it float the longest time possible. The winner will be whoever keeps the ring floating the longest.

ACTIVITIES

24. Rub a ruler and a plastic case with wool, and a comb with hair. We observe that the ruler and the case repel each other when they are put together, whereas the comb and the case attract each other. Explain these observations with the help of your partner. What will happen if we put the ruler and the comb together? (The behaviour may be different depending on the materials.)

25. Sometimes when we get out of a car that has been driven and we come near the car body we feel a shock. Think with your partner: why do you think is the reason for this and why is there a second shock? What can you do to avoid these shocks?

6 Magnetic force

You know about electric force and gravitational force, but what interaction do magnets have? We will explain it with some experiments.

PHYSICS 2.0

Magnets create what we call a **magnetic field** around them. The following simulation will help you to visualise this and to observe how it affects the compass.

goo.gl/pmDnQY

You can also see how the Earth's magnetic field behaves and how its poles are distributed.

CLASSROOM LABORATORY **Playing with magnets**

4. In pairs, do the following experiments and photograph what happens. Make a presentation with all the results and explain what you have observed and present it to your class:

   a) Take two magnets and a cardboard tube from the inside of a roll of kitchen paper. Cut the tube in half lengthwise and put the magnets inside facing each other. What happens? Now turn one around. Has something changed?

   b) Get a magnet, some keys, an empty fizzy drink can, some copper, gold and silver rings and any other metal object you can find. Put the magnet near each of them and see if they attract each other. Make a table to record the results.

   c) Take a transparent plastic or glass bottle. Put iron filings or small pieces of iron wool in the bottle and fill it with sunflower or body oil, shake it and cover it. Put a magnet near the bottle and move it around.

The experiments with magnets show that the force that appears between them fulfils the following characteristics:

- It acts at a distance and decreases when the distance increases.
- It can be attractive or repulsive.
- The magnet has two differentiated parts called the **north poles** (N) and **south poles** (S). Like poles repel and unlike poles attract.
- It only affects certain metals like iron.

REMEMBER

Steel is an alloy or homogeneous mixture of iron and carbon.

This is why it is sensitive to the magnetic interaction. It is useful for making artificial magnets that are more resistant than natural ones or those made only with iron.

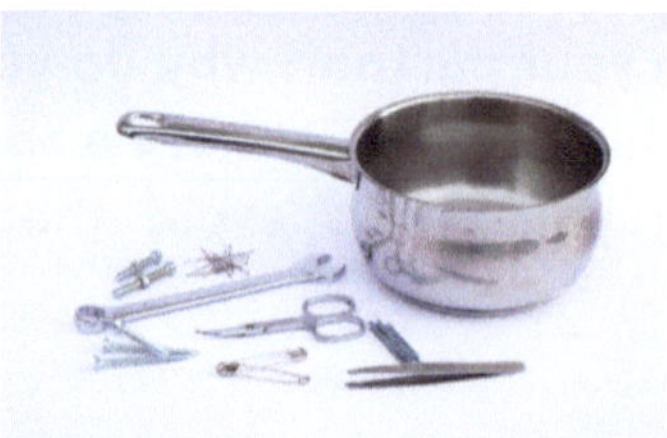

6.1. Similarities and differences with electric force

Although they are similar, electric force and magnetic force are different:

They are similar	They are different	
	Electric force	**Magnetic force**
They act at a distance.	There are bodies with only one charge.	We cannot separate the poles; they always appear in pairs.
There are two types of charges (+ , –) or two types of poles (N, S).	The charges can move through matter.	The poles cannot move through matter.

7 Electricity and magnetism

Experiments in the nineteenth century demonstrated that electricity and magnetism are aspects of the same property, **electromagnetism**.

WORKSHEET

Download this unit's worksheet and do activities 22, 23 and 24.

EXPERIMENT The compass

3. Use the compass that you will build in the laboratory at the end of the unit or a shop-bought one. You will see that even though you spin the needle, it always marks north. Now put it next to a transformer like a mobile phone charger or another device. What do you observe?

   Move the compass in a circle around the charger and describe what happened. Why do you think the needle acts like this? Experiment with different chargers.

The behaviour you observed allows us to interconvert between both properties. This has been one of the greatest advances of humankind. It gave us a simple and cheap way of obtaining, transporting and using energy.

Magnetism → Electric current

Get a bicycle dynamo and take it apart. You will see some magnets and inside there is a copper coil.

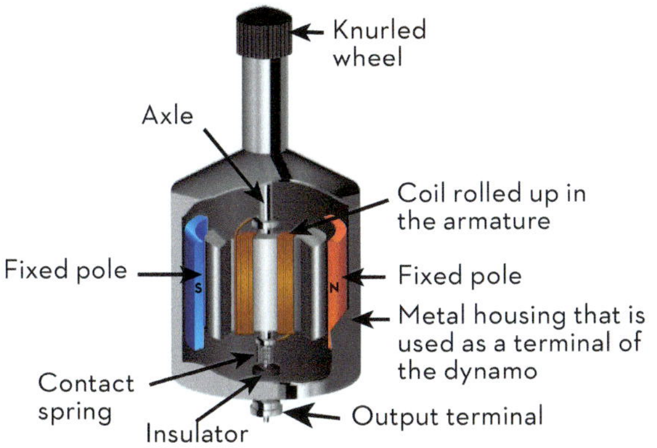

When the copper coil turns inside the magnetic field a current is induced in the coil. It can be used for lighting.

Electric current → Magnetism

An electric current creates an area of influence around itself called a magnetic field and it behaves like a magnet.

When the current stops, it stops behaving this way.

ACTIVITIES

26. Induction cookers are based on creating a variable magnetic field on the pan. This causes an electric current in the base of the pan, causing it to heat up. You can differentiate if a pan is for induction or glass ceramic cookers if a magnet sticks to its base. Why? Discuss your answer with your partner.

27. If all packaging is thrown into the same yellow recycling bin, what procedure is used to separate the cans from the other waste?

MIND MAP

Download and print the worksheet you will find in the OLC to complete the mind map of the unit.

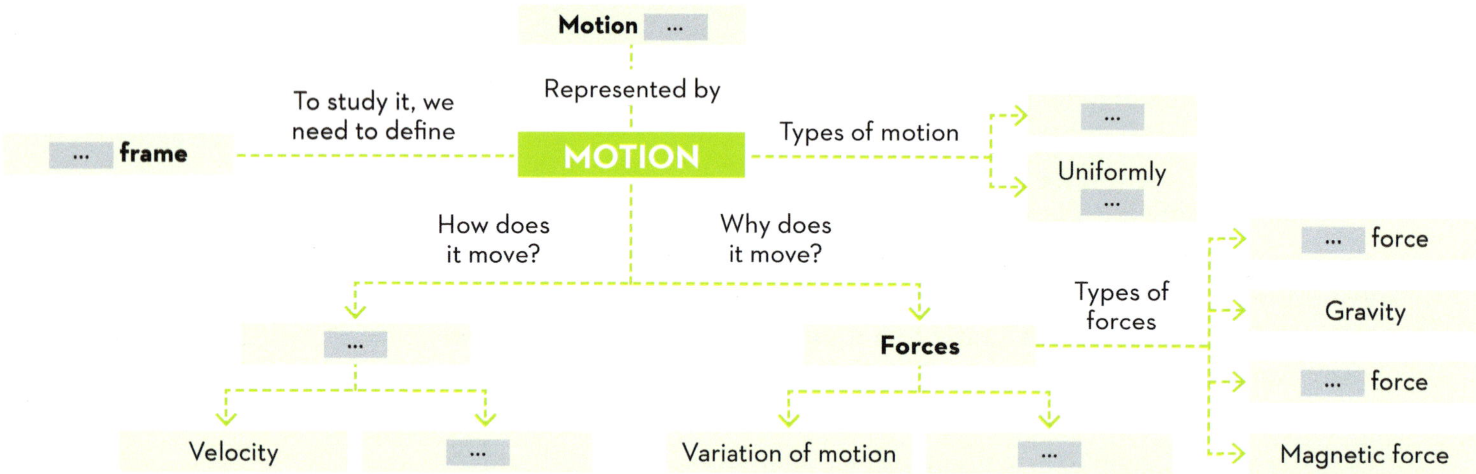

LOOK AROUND YOU

The *Mars Climate* crashed on Mars because NASA did not translate kilometres to miles

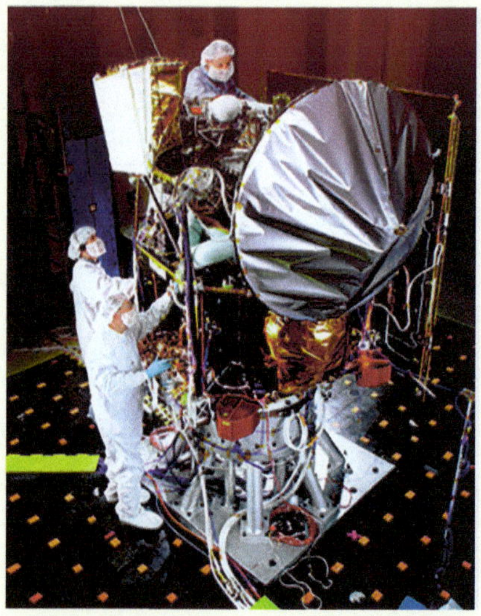

For some time now American public institutions, from the CIA to NASA, to the White House and the Pentagon, are not even perfect in Hollywood films. But sometimes their errors are almost ridiculous. This is the case of the *Mars Climate Orbiter*, which crashed into Mars last week. According to NASA reports, the failure came from confusion between miles and kilometres. As simple as that. The probe, built to navigate according to the English system, received flight instructions before taking off in the decimal metric system.

The *Jet Propulsion Laboratory* in Pasadena, in charge of programming the probe's navigation systems, uses the decimal metric system (millimetres, metres and kilometres) to make its calculations, whereas the other laboratory, Lockheed Martin Astronautics in Denver, which designed and built the *Mars Climate Observer*, uses the English system (inches and feet).

However, the navigation data were not converted from one system to the other before launching the *Mars Climate* into space. Consequently, the ship suffered severe confusion, a kind of schizophrenia which led it to reach the red planet in the wrong orbiting position, making it crash. Last 23 September, the artefact was lost and now must be complete space junk. Junk that cost American taxpayers a chilling 125 million dollars. NASA's report recognises this foolish error with embarrassment [...]

Elpais.com. 02/10/1999 (adaptation)

Questions

With the help of your partner, answer the following questions:

a) What was the error that NASA committed?

b) Why can't the ship be stopped from the Earth?

c) If we drove a car at 100 mph, would we break the legal speed limit in Spain?

LABORATORY WORK

Build a compass and find your way

Objective

Nowadays we have a multitude of electronic systems which help us find where we are on a map, so we can get to the right place. But until not long ago this was not possible. Travellers needed another instrument that was essential for worldwide trading: the **compass**.

You might even need it. What would happen in inhospitable places if your mobile phone battery ran out?

Material

You need: a pin, a magnet, a small polystyrene disc and a container with water.

Procedure

Working with your partner, follow the next steps:

First you need to get a very light magnet:

- Successively move the pin over the magnet, **always** in the same direction. The more times you repeat this, the longer and clearer its effect will be.

  If you do not have a magnet, you can also point it approximately toward North (away from the Sun) and hit it hard with a rock, or better, with a hammer. In this case, the effect will be more difficult to observe, but it still becomes a magnet. Check this by putting it near an iron object.

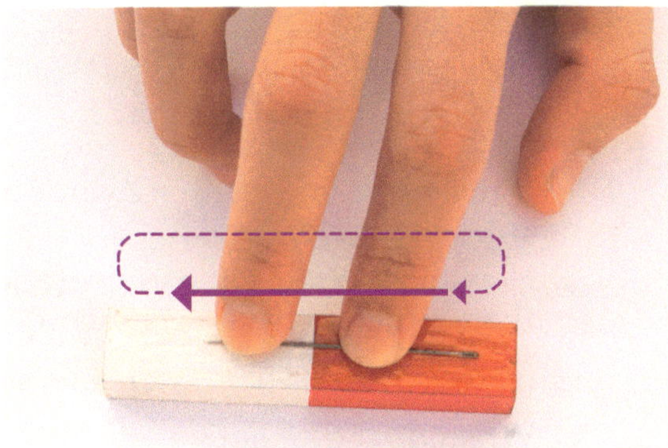

The second step is to give it free movement.

- Put it on the polystyrene disc and the disc in the container with water, as shown in the picture.

Check the result:

- If you let it go, you will always see the needle point toward the same direction. Depending on how you moved it over the magnet, the tip will always point to north or south. To identify them, if you are in the Northern Hemisphere, the end of the needle that is closest to the Sun is the south.
- Now put a second magnet or even a metal object near it and you will see how your compass will point to it.

Task

Take photographs of the whole process of making the compass and the results obtained. With the help of your partner, exhibit your work on a poster, explaining the process you followed and the consequences you observed.

FINAL ACTIVITIES

Basic activities

1. Working in pairs, identify the movables, the reference frame, the positions and the displacement in the following text.

Daniel and Marcos are returning to their house, which is three kilometres from the bus stop. Daniel gets tired of waiting for the bus and starts walking at 9 o'clock. In ten minutes the bus picks up Marcos and five minutes later they pass Daniel at the fountain along the way.

At twenty past nine, Marcos arrives home and sits to wait for his brother, who arrives at half past nine.

Make a graph that shows the displacements and determine the velocities of each one. Where is the fountain located?

2. Two movables move as shown in the graph below. Determine:

a) How many different motions does each of them experience? Name them.

b) Determine the velocity of each section.

c) When do they meet?

d) What displacement has each one of the bodies experienced in total?

e) Discuss your answers in small groups.

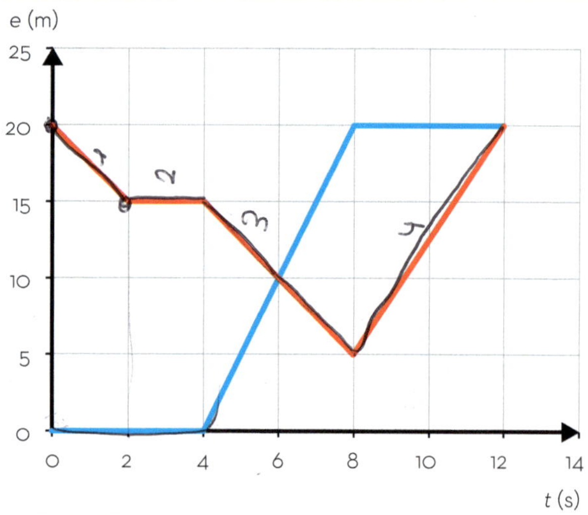

3. Determine the accelerations of each of the sections of the following graph.

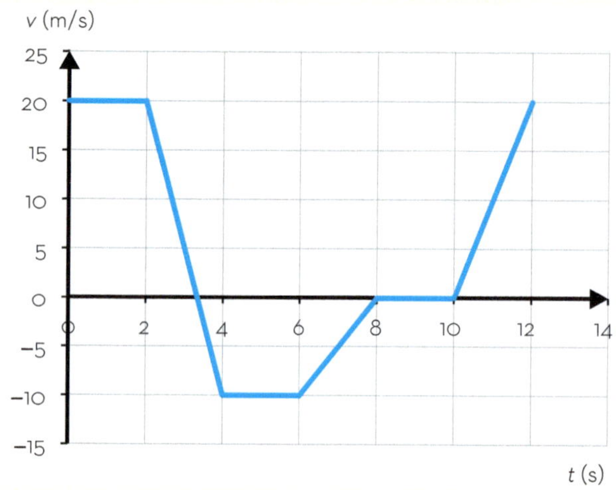

4. A car that is moving at 90 km/h needs to pass a lorry that is moving at the same speed, so it needs to reach 120 km/h in 3 seconds. If the car has an acceleration of 1.6 m/s², can it do it? How long does it really take?

5. How many ways can forces act and what may be their effects? Indicate two examples in each case and share them with your classmates.

6. How do bodies behave under the forces of deformation? Identify two examples of each case in your home and take photographs to show this behaviour. Explain your examples to the rest of the class in a small presentation.

7. Determine the total force that acts on each body, as well as the acceleration that is observed:

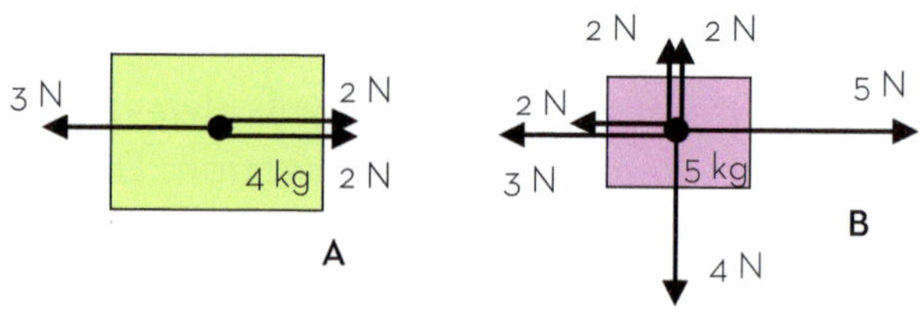

8. A friend tells you she weighs 60 kilograms. What comment would you make (in regard to physics) and what is her weight on the Moon?

9. Uranus is $2.87 \cdot 10^9$ km from the Sun. How long does it take sunlight to reach it?

10. Identify three situations where there is static electricity in your home. Take photographs of each situation and make a small presentation to show to your class.

Consolidation activities

11. A moveable moves as shown in the graph below. Determine:

a) How many different motions does it experience? Name them.

b) Determine the velocity of each section.

c) Determine the average velocity between second 5 and 15.

d) Make the velocity-time graph.

e) Compare your answers with your partner.

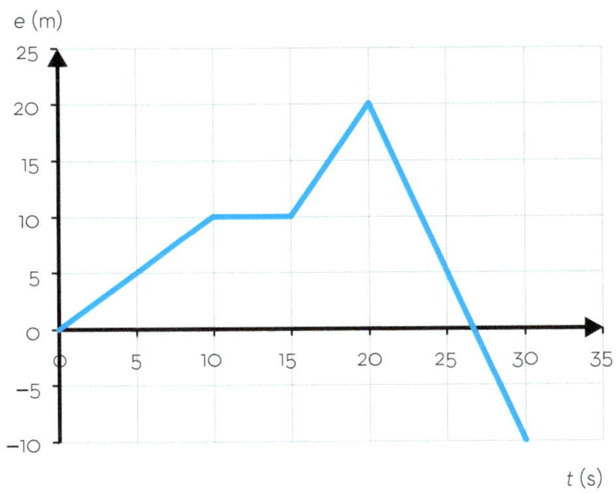

12. A man pushes a shopping trolley with the data from the figure below. Draw the forces. What is the acceleration that the trolley acquires? What velocity does it have in 5 s?

Mass = 50 kg

F_{Man} = 1250 N

$F_{Friction}$ = 150 N

It starts from a resting position.

13. You weigh a rock on Io, one of Jupiter's moons and you obtain the value 36.2 N. You bring it to Earth and it weighs 196 N. What gravity does Io have?

Advanced activities

14. Work with your partner: a car takes 24 minutes to go from city A to city B. What velocity does it have?

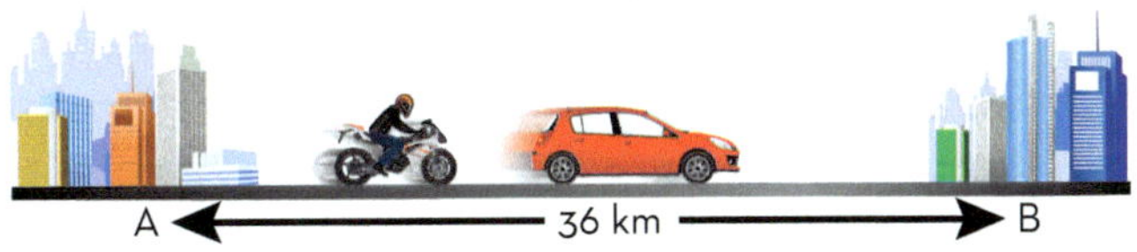

If a motorcycle starts the same journey four minutes after the car left, what must its minimum velocity be to arrive before the car?

Show both motions on a graph.

15. The spring constant of a spring is 150 N/m. Working in pairs, answer the following questions:

a) What force do we have to apply to stretch it 30 cm? What about to compress it 30 cm?

b) What is its elongation if we apply a force of 25 N?

16. Calculate the value of the friction that a 15 kg body has. When we push it with 300 N, we observe that it acquires an acceleration of 15 m/s^2.

17. What is the distance of Io from the Sun if sunlight takes 43 minutes and 14 seconds to reach it? Compare your answer with your partner.

18. We put a body at a distance of 300 km from the Earth's surface. What do we have to do so that it does not fall to Earth? Discuss your answer with the rest of the class.

19. You probably have heard of wireless chargers. These are common in some wireless telephones and other small devices.

Research how they work. Base your research on the conversion between electricity and magnetism. Record a video with your explanation and show it to your class.

The kinetic drop

The study of braking conditions is of great importance for educating drivers and preventing accidents.

In 2011, the Spanish Association of Traffic Accident Investigators (AEIAT – Asociación Española de Investigadores de Accidentes de Tráfico) analysed the influence on the braking distance of a dry surface or a wet surface at different distances (**goo.gl/jAeaE6**).

The report shows the different braking distances at two driving velocities on different surfaces. The result can be summarised in the following graph.

	Velocity	Reaction distance	System reaction	Braking distance	Stopping distance
Wet	100 km/h	27.8	6.8	90.2	124.8
Wet	90 km/h	25	6.1	72.8	103,9
Dry	100 km/h	27.8	6.8	47	81.6
Dry	90 km/h	25	6.1	37.8	68.9

Distances in metres.

With this data, answer the folowing questions. Then check your answers with your partner:

Question 1

At the moment when the car starts to brake, which of the following options is correct?

A. Acceleration is zero, so the car stops.

B. Velocity is zero, so the car stops.

C. Acceleration is negative, so the car stops.

D. Velocity is negative, so the car stops.

Question 2

What was the cause of the car stopping in the experiment?

A. Because no force was applied to the car.

B. Because there was a force of friction that opposes motion.

C. Because there is a force of friction in favour of motion that prevents the car from stopping instantly.

D. None of the above.

Question 3

When the experiment is conducted on a wet surface, why do we observe that the braking distance increases?

A. Because there is no acceleration on water.

B. Because there is no velocity on water.

C. Because there is less friction on water.

D. Because the wheels keep turning on water.

Question 4

Which of the following expressions describes the motion of the vehicle at 100 km/h on a wet surface?

A. Velocity starts to decrease when the reaction distance starts.

B. Velocity starts to decrease when the system reaction starts.

C. Velocity starts to decrease when the braking distance starts.

D. None of the above.

Don't let it break!

→| Competency task

Objective

The objective of this task is very simple: you have to prevent a balloon full of water, thrown from the second storey of a building, from breaking when it falls to the ground. Simple, right?

Final product

You must record the whole design process, as well as its implementation and produce a video, like a documentary where you show what you have done. You must add the physical reasons it is based on, using the concepts of forces, velocity and acceleration.

Step by step

You must do the work in groups with a maximum of four members. Although you can choose your own way, we suggest you follow a series of steps:

1. Look up information on different procedures that other people have done. You are not going to copy them, but rather obtain ideas for designing your own system. Originality will be rewarded in the work.
2. Get together and brainstorm. Write down all the possibilities that occur to you, no matter how improbable they seem.

3. From all the ideas, select the two easiest because they have more possibilities for success.
4. Now go on to the practical part and try to achieve your goal. Start with balloons with a small amount of water.
5. Analyse up to what point you can increase the mass of the balloon and the system you chose is still able to achieve your goal.

Self-assessment

Evaluate your work and your classmates' work with the guidelines that you can download from the OLC.

1st STAGE: IDENTIFYING THE PROBLEM

We have seen that **electricity** and **magnetism** are closely related. This fact was discovered in two of the most important experiments of the early 19th century.

In this investigation, you are going to replicate these experiments. The objective is to gain as much information as possible in order to find applications for the new phenomena.

The problem is:

WHAT IS THE RELATIONSHIP BETWEEN ELECTRICITY AND MAGNETISM?

2nd STAGE: FORMULATING THE HYPOTHESIS

The research topic is to find the connection between electricity and magnetism. It is time to identify possible relationships. To do this, write down the possible dependencies between the two phenomena with the help of your partner. Use the words in the box below and follow the example to write your hypothesis. For example: *A magnet resting near a cable creates an electric current.*

Remember, these are your ideas. You can check whether they are true or false later on.

create, magnet, electric current, magnetic field, moving, resting, constant, variable, coil

3rd STAGE: RESEARCH

You are going to do two experiments: Oersted's experiment and Faraday's experiment. Each one consists of two investigations. Do the experiments with the help of your partner and answer the questions together.

Oersted's experiment

First investigation

For this experiment you will need: a battery (preferably a 4.5 V square battery), a switch, an insulated conductor wire (enamelled or plastic-coated) and a compass (remember, you made one in the laboratory work in Unit 6).

Assemble the circuit in the diagram:

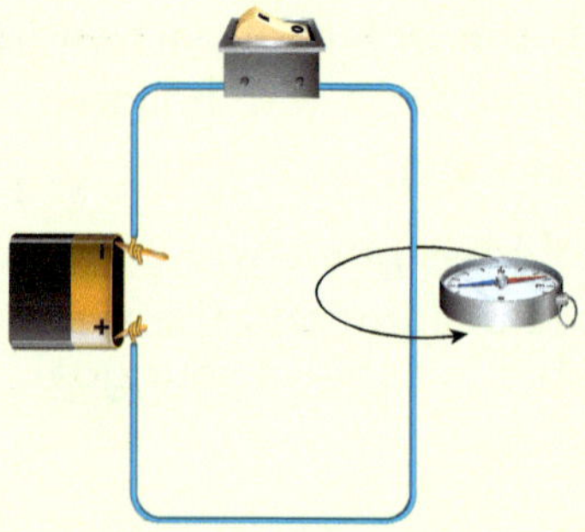

Do not keep the current on for long or it will ruin the battery.

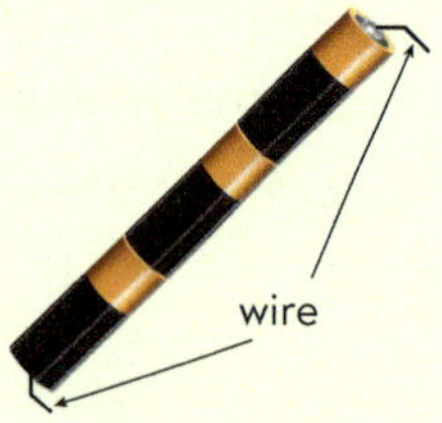

If you do not have a square battery, use three normal batteries joined together.

Keep the switch closed and move the compass around the wire, as indicated in the diagram.

What is making the needle move?

Second investigation

With the same circuit, take the wire and wind it around a nail.

- Reconnect the wire. Now put the compass near it again.
  What do you observe when you move the compass around the wire?
- Put a paper clip near the nail.
  What happens?
- Open and close the switch.
  What happens?

Faraday's experiment

First investigation

Material: insulated wire, two LEDs or a 2 V bulb, a magnet and a cardboard tube.

Working in pairs, assemble the circuit in the diagram and do the following:

- Put the magnet inside the cardboard tube.

  What happens?

- Now remove the magnet.

  Does the same LED light up?

- Repeat the experiment, but reverse the magnet.

  What changes do you observe? Does the same thing happen if we move the magnet faster?

Join the short terminal of one LED to the long terminal of the other LED.

Wind the wire twenty times around the cardboard tube from a roll of kitchen paper.

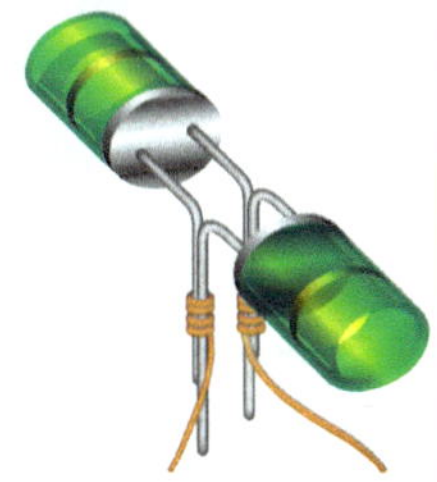

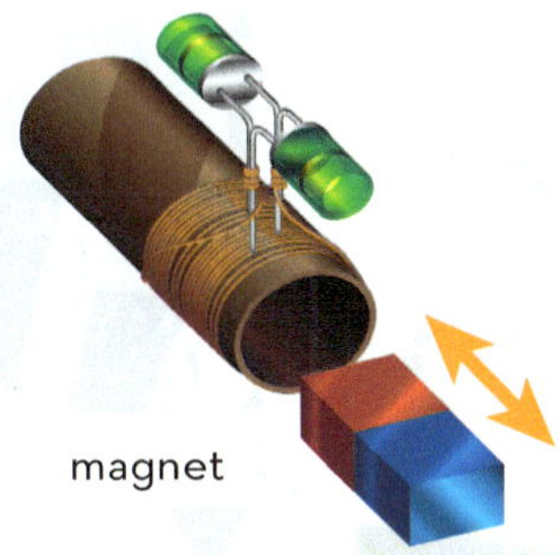

Note: Hardware shops sell powerful magnets, but you should handle them with an adult present. You can also use magnets from the game *Geomag*.

Second investigation

- Using the same circuit, leave the magnet in the same position, but move the tube.

  How do the LEDs respond to the tube's movement?

- Now invert the magnet.

  Do the same LEDs light up?

  Is the result of moving the tube the same as that of moving the magnet?

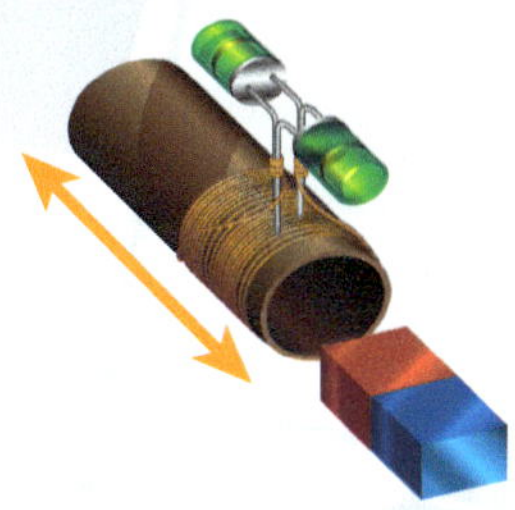

4th STAGE: MATERIALS AND METHODS

Working with your partner, write a report about your investigation. Follow these steps:

- Give a brief **introduction** to your research: the subject and the experiments.
- **Describe** the experiments you have conducted. Use photographs and include your answers to the questions.
- These experiments have applications in instruments you have around you. **Find them**, take photographs and comment on them.

5th STAGE: CONCLUSION

Look at your hypotheses from the second stage. Were they correct?

Repeat Faraday's experiment winding the wire five times and then fifty times.

Conclude your project with your opinion on the usefulness of the relationship between electricity and magnetism.

You must present your project in two ways:

- Write a report on all the information.
- Together with your partner, record a video showing your research to the rest of the class.

7 ENERGY

In the history of mankind, three important events have completely changed how we live. The first was the discovery of **fire** and how to manipulate it. This helped humans to warm themselves in winter, it provided protection against animals, and it changed their diet.

The second change occurred around 1750: we call it the Industrial Revolution. The most significant discovery was how to transform **steam** into motion. This is linked to the use of machinery, different types of transport and industry.

The third transformation came at the end of the 19th century with the discovery of **electricity**. It radically transformed people's lives and provided homes with different types of energy: heat, motion, light and sound.

Fire, steam and electricity are three manifestations of **energy**, the main concept of this unit. Energy is one of the key ideas in the history of science and mankind. Humans need abundant and safe energy sources. It is also important that these energy sources do not endanger the future of our planet. Nowadays there is lots of research into new sources of energy. **Renewable energy** sources are slowly replacing fossil fuels such as oil and coal. There is also research into using the same energy as the stars use: nuclear fusion.

> 'Humanity stands before a great problem of finding new raw materials and new sources of energy that shall never become exhausted. In the meantime we must not waste what we have, but must leave as much as possible for coming generations.'
>
> Svante Arrhenius, Nobel Prize for Chemistry

Contents

WORK WITH THE TEXT

Answer the following questions about the text:

1. Name the three discoveries that have changed how we live.
2. What did humans use fire for in prehistory?
3. Do you know how a steam machine works? Try to explain it.
4. Working with your partner, explain the meaning of 'renewable energy source' and give an example.

CHALLENGE

The batteries in our mobile phones accumulate energy and transform it into light, sound, waves, images and colour.

Turn off your mobile and take out the battery. Find the battery capacity (for example, 2600 mAh) and the nominal voltage (for example, 3.8 V). Write down the values and do the following calculation:

Battery energy (J) = battery capacity (mAh) · 3.6 · voltage (V) in this equation stands for joules, the unit we use to measure energy.

In your mobile, go to *Settings/Battery*. Look at the percentages that show how your phone consumes the energy in its battery. Now discuss with your partner:

- Which one of you has the best battery?
- How much energy (in joules) do your batteries still have?
- Why do your mobile batteries run down?
- Which systems on your mobiles consume the most energy?
- Suggest ideas about how to reduce this percentage.

KEYWORDS

Download and print the activities that will guide you through the essential vocabulary of this unit.

FIND A CLASSMATE

Download and print the worksheet 'Find a classmate: energy sources'. Share experiences about this topic with your classmates and discover interesting things about them.

1 Energy

Physics and chemistry are sciences that study the behaviour of matter. Anything that happens to a material object implies a change in this object or its environment. Science has a very powerful tool to study these changes: **energy.**

DID YOU KNOW?

When we heat a sheet of steel, part of its energy becomes luminous radiation. At a low temperature it is red, but it can become white or even blue as the temperature increases. The same phenomenon occurs with stars: we can know their temperature according to the colour we see.

THINK AND REASON

Which of these systems has a greater capacity to produce a change in its surroundings? What types of changes can they produce?

- A red-hot sheet of steel or the same sheet at room temperature.
- A car travelling at 100 km/h or the same car travelling at 60 km/h.
- A kilogram of coal or a kilogram of wood.

WORKSHEET

Download this unit's worksheet and do activity 1.

Scientists analyse the changes that occur in nature. To predict these changes, they defined a magnitude called energy. It is one of the most important magnitudes in science.

Energy is the magnitude which quantifies the capacity of an object to produce changes. These changes can be within itself or in its environment.

As you learned in the unit on scientific activity, a magnitude is any property of a system that we can measure. This means we can give it a numerical value and a unit.

The **unit** of energy in the International System is the **joule (J).**

Other units of energy are the calorie (cal), the kilowatt hour (kWh) and the tonne of oil equivalent (toe).

$1\ \text{cal} = 4.19\ \text{J}$ $1\ \text{kWh} = 3.6 \cdot 10^6\ \text{J}$ $1\ \text{toe} = 4.19 \cdot 10^{19}\ \text{J}$

ACTIVITIES

1. **What type of changes can a litre of petrol, a car travelling at 80 km/h and a mobile battery produce? Assign the following quantities of energy to each one: $3.5 \cdot 10^4$ J, $3.5 \cdot 10^7$ J, $3.5 \cdot 10^5$ J.**
2. **Interpret the graph using the definition of energy given above. Comment on your results with your partner.**
3. **What needs more batteries: a talking doll or a doll that talks and crawls?**

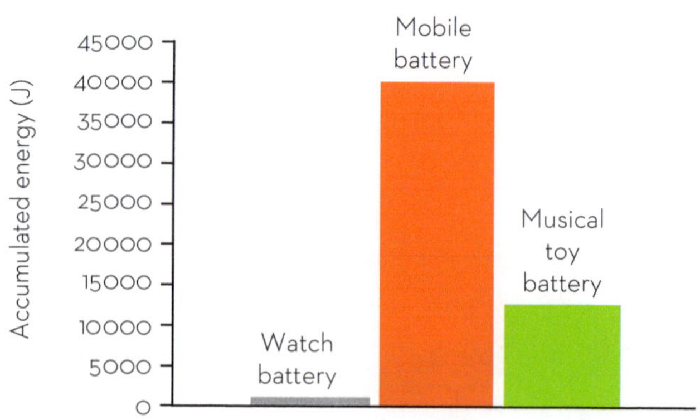

1.1. Types of energy

Objects can produce changes in many ways. Scientists group these properties into different types of energy. Observing a family in their country home, we find many types of energy.

1.2. The properties of energy

An object has a specific amount of energy depending on its **velocity**, **height**, **temperature** and **molecular properties**. This amount of energy can:

- **Transform** into another type of energy that still belongs to the same object. For example, a television transforms electrical energy into electromagnetic energy.
- **Transfer** to another object in the same form or in a different form. For example, the chemical energy of burning wood transforms into thermal energy. This thermal energy is then transferred to the receptacle on the fire.
- **Dissipate.** This means that the energy is transferred to the air, to water or to the ground. It is divided among all these molecules, so the energy becomes useless. For example, part of the thermal energy from the fire dissipates into the surrounding air.

The most important thing to remember about energy is:

Energy is always conserved. It cannot be created or destroyed. Any process always obeys the following rule:

Initial energy = Transformed energy + Transferred energy + Dissipated energy

ENERGY 2.0

Go to the following link:

goo.gl/Ttda46

Combine different elements to see how energy is transformed, transfers and dissipates, but it is always conserved.

WORKSHEET

Download this unit's worksheet and do activities 2 and 3.

WORKSHEET

Download this unit's worksheet and do activities 4, 5 and 6.

Energy is conserved and this is its main property. If we know the energy an object has, we can calculate the magnitude of changes this object can produce. This is very useful for science and technology. Here are two examples to help you understand the properties of energy.

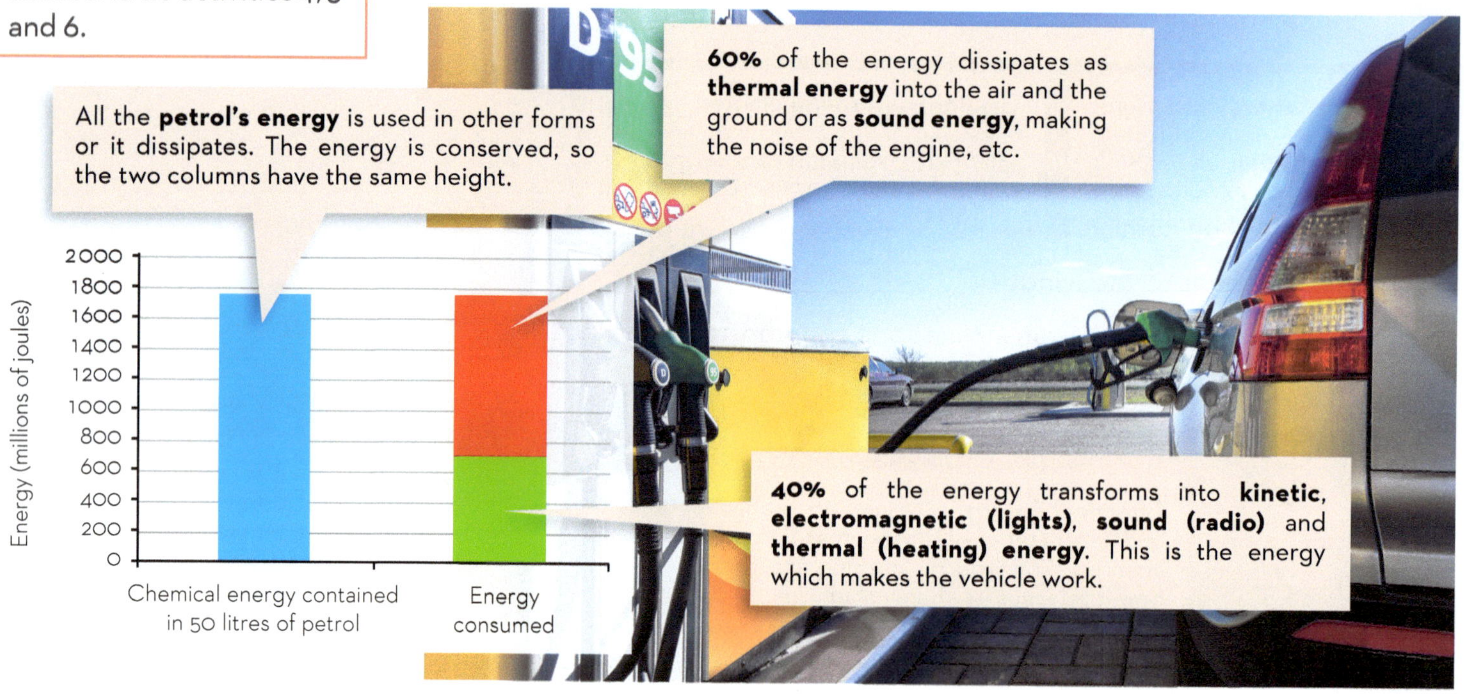

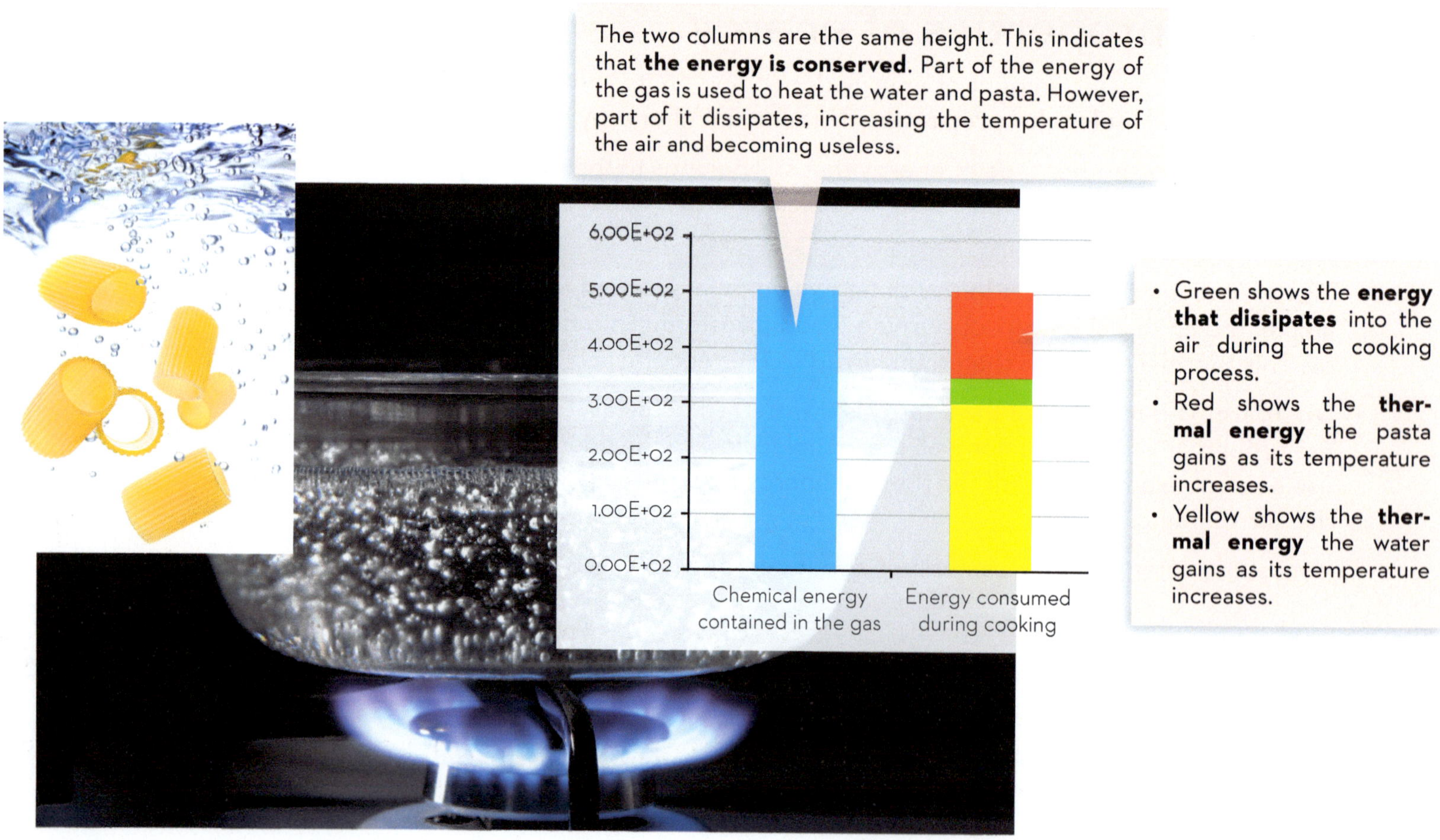

CLASSROOM LABORATORY Different types of energy

1. Any laboratory experiment produces changes and therefore involves different types of energy. For the following experiment you will need: a system to heat water (hotplate, Bunsen burner, etc.), a saucepan, water, a funnel and a paper windmill. (You can find instructions on how to make the paper windmill at **goo.gl/nbceeW**.)

   Put the water in the saucepan and bring it to the boil. Carefully hold the funnel upside down (with the wide part at the bottom) over the steam. Put the windmill above the narrow part of the funnel, where steam should be coming out.

   - What happens to the windmill?
   - Indicate all the energy transformations and transferences that occur in this experiment. Is energy dissipating in any way?
   - If we turn off the heat, the windmill will eventually stop spinning. Explain why using the concept of energy conservation.

EXPERIMENT

1. Look for a semi-spherical object at home, such as a salad bowl, cereal bowl, etc. Put a marble on the top part and let it fall into the receptacle.
   - What type of energy does the marble have when you let it rest on the top part of your semi-spherical object? What type of energy does it have when it finally stops? Is there conservation of energy?

ACTIVITIES

4. Create a poster similar to the compositions in section 1.2. Find three images on the internet and describe the energy transformations linked to each image. Make sure you use examples involving the transformation, transference, dissipation and conservation of energy.

5. Working with your partner, describe the energy transformations that take place in the following situations:

The energy captured by a solar panel transforms into the energy transmitted by a television.

The electrical energy that heats a ceramic hob results in a cooked potato.

The sun's electromagnetic energy results in a sack of flour.

The energy from a sack of flour results in a person running in the park.

6. Laura is four years old and has a toy phone with batteries. The keys light up and make sounds when pressed. Laura does not understand why her father says that the batteries will run out. Can you explain why?

2 Energy, heat and temperature

ENERGY 2.0

To understand the concepts of temperature and thermal energy, watch the simulation at

goo.gl/qm4h1r

When microwaves transfer thermal energy to the molecules, they increase the kinetic energy of the molecules. Therefore, the temperature of the water also rises.

WORKSHEET

Download this unit's worksheet and do activities 7 and 8.

In the unit on the properties of matter, you learned that the temperature of an object is a consequence of the average speed of its particles. These particles can produce changes because of their speed. Therefore, they have a type of energy called **kinetic energy.**

The **temperature** of an object is a property which measures the average kinetic energy of the particles in the object.

The capacity of objects to produce changes because of their temperature and their mass is called **thermal energy**. Therefore, thermal energy is a macroscopic term to measure the kinetic energy of the enormous number of particles in an object.

The higher the temperature of an object, the more kinetic energy its particles have. The greater the mass of an object, the more particles it has. Therefore, the higher the temperature and the greater the mass, the more thermal energy an object has.

Heat is the term used in physics to group all the ways of transferring thermal energy between objects. Objects have **thermal energy** and this transfers to other objects as heat. This transference always goes from the object with the higher temperature to the object with the lower temperature.

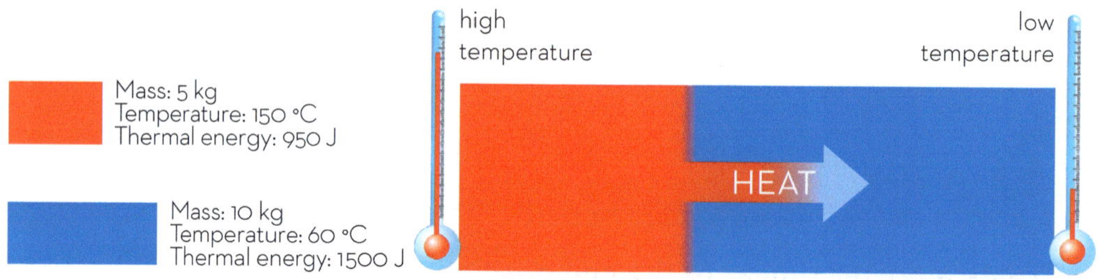

ACTIVITIES

7. Which has more thermal energy: a bucket with 20 litres of water at 60 °C or a glass full of water at 80 °C? What experiment could you do to prove this? Is there a transfer of energy as heat in this experiment? Share your results with your partner.

8. Imagine we put the water from the two receptacles in Activity 7 together. Would the energy transfer as heat from the water at 60 °C to the water at 80 °C or vice versa?

9. Sparklers burn at temperatures of over 1000 °C. Working in pairs, apply the ideas from Activity 7 to explain why the small sparks from a sparkler are not dangerous.

10. Compare the normal meaning of the word 'hot' with the scientific meaning based on the following phrases: 'I'm so hot!' and 'The soup is very hot.' Write them in a way that is scientifically correct. Compare your sentences with the rest of the class.

2.1. Ways of transferring thermal energy

There are three ways an object can transfer thermal energy to another object:

CONDUCTION

Conduction is the transference of thermal energy between two objects as a result of collisions between their particles. In these collisions, the particles of the object with a higher temperature transfer part of their energy to the object with a lower temperature. It is the same as when two balls collide: the faster ball increases the speed of the slower ball.

Your hand or the air heat up by conduction when they are in contact with the radiator.

Energy transfers from the area with a higher temperature to the area with a lower temperature.

If two objects are in contact for enough time, their temperature ends up being the same. The transfer of energy through particle collision then stops because their particles have the same amount of energy. At this point we say the objects have reached **thermal equilibrium.**

The redistribution of energy by conduction depends on the characteristics of the colliding particles.

- Materials such as metal are good at transferring heat. They are called **thermal conductors**.
- Materials such as plastic are not good at transferring heat. They are called **insulators**.

CONVECTION

Convection is the transference of thermal energy linked to the movement of matter. It occurs in liquids and gases. These are the states of matter which can flow.

The air in contact with a radiator increases its temperature by conduction. As the air temperature increases, its density decreases and the now hot air rises. Cold air is denser, so it descends and comes into contact with the radiator, increasing its temperature. This is how all the air in the room heats up.

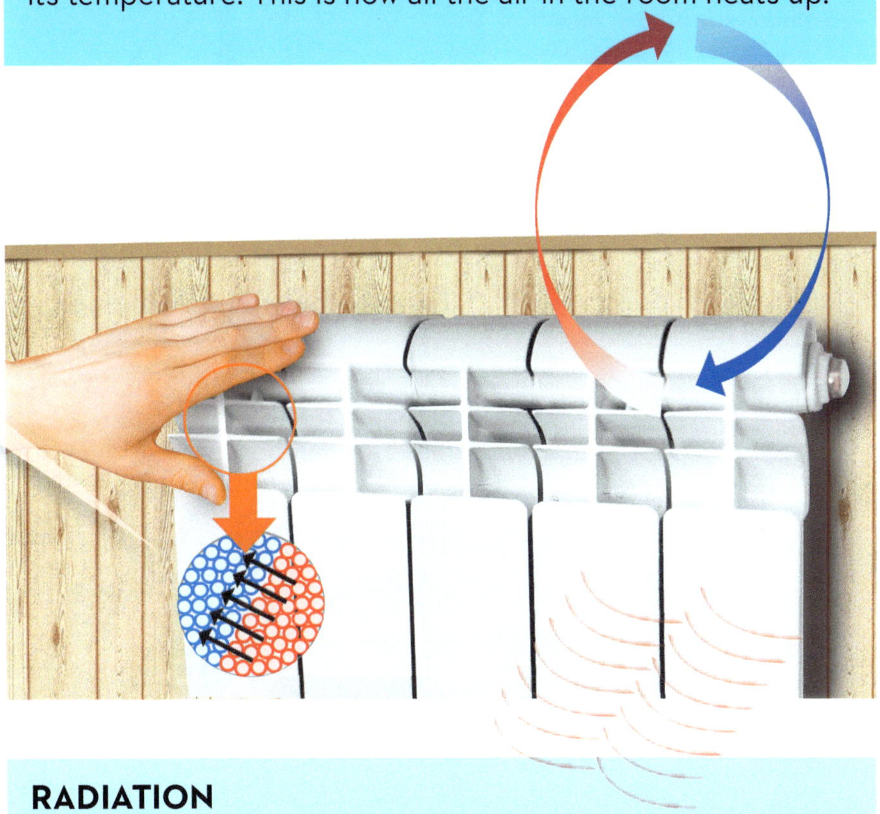

RADIATION

Radiation is the transference of thermal energy through electromagnetic waves. Radiation is the only mechanism which can occur in a vacuum. In other words, where there is an absence of matter. All objects emit energy by radiation. The higher their temperature, the more energy they emit.

The particles in the metal of the radiator are very hot, so they vibrate with a lot of energy. They emit part of this energy in the form of waves. These waves transfer the energy to any other object they come into contact with.

Conduction, convection and radiation are present everywhere in our daily lives. Here are some examples.

The energy from the sun reaches Earth by **radiation**. In other words, in the form of electromagnetic waves. Remember that there is no matter in space. Therefore, this is the only possible way this energy can be transferred.

The water vapour that evaporates from the sea rises by **convection**. When it reaches a certain height, the temperature is much lower. The water then condenses to form clouds.

When we swim in the sea, we notice that the deeper water is colder. Cold water has a higher density, so it sinks downwards. The surface water receives the sun's energy and warms up. Because it is less dense, it stays at the top.

Air is an excellent thermal insulator. That is why double-glazed windows are very effective. The two panes of glass have to be close together to stop convection currents from forming. Convection currents transfer thermal energy from one place to another.

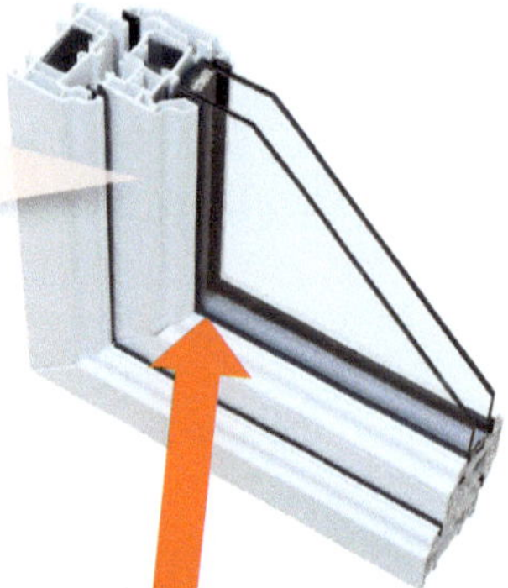

Wind is the movement of air. The main factor in wind is **convection** currents. Hot air rises and a cold mass of air takes it place, producing wind.

WORKSHEET

Download this unit's worksheet and do activities 9 and 10.

In tall buildings, the higher flats are warmer than the lower flats because of **convection** currents.

Plastic is a thermal insulator. The plastic handle stops the thermal energy from the pan reaching your hand.

They are frying vegetables in this kitchen. The pan is metal, an excellent thermal conductor. This allows the thermal energy from the flame to reach the vegetables by **conduction**.

Modern houses have thermal insulating material inside their walls. This stops heat being transferred from inside to outside the house through **conduction**. Old houses did not have this material, but they had thicker walls. This made the exchange of heat difficult.

2.2. Energy transfer by sound

Sound is a form of energy transference through sound waves. In contrast to electromagnetic waves, sound waves cannot travel in a vacuum. This is because their energy is propagated through the vibrations of physical matter.

A good example is a guitar string or the membrane of a drum, where kinetic energy transforms into sound energy.

WORKSHEET

Before experiment 3, download this unit's worksheet and do activity 11.

EXPERIMENT

How do you wrap up warm?

2. Wrap one ice cube in aluminium foil and another one in a jumper. Wait 45 minutes, then unwrap the ice cubes and observe what has happened.

   Which material is a better thermal conductor: the aluminium foil or the fabric?

   A pupil makes this comment: 'Jumpers keep us warm in winter, but this one kept the ice cube cold.' Explain this apparent contradiction to the student.

Dancing ink

3. We are going to observe the phenomenon of convection using hot and cold water. You will need: water, ink or colorant for the water, a small glass jar (a jam or pepper jar, for instance), and a large glass jar or a plastic bottle. Make a hole in the lid of the small jar (using a screwdriver and a hammer, for example). Fill the large jar or bottle with cold water with no colorant. Now fill the small jar with coloured hot water and put it inside the large receptacle filled with clear cold water.

   What happens to the hot water in the small jar? Explain this behaviour.

ACTIVITIES

11. People have compared heat conduction to a human chain that passes a bucket of water from person to person. The people do not move, but the water does. Can you explain this comparison?
12. Find information about a system called underfloor heating. Why is it installed under the floors in houses? Why are air conditioning vents installed in high locations? Discuss your answers with your partner and after reaching an agreement share your answer with the whole class.
13. Greenhouses are essential to farming nowadays. Make a slide show about these constructions. What are they for? What concepts of physics do they use? What is planted in them? Give an oral presentation of the slide show in class.
14. Make a poster to explain how the human ear works. In particular, show the transference of sound to mechanical energy that occurs in our eardrum.

3 Thermal expansion

WORKSHEET

Download this unit's worksheet and do activities 12 and 13.

When an object absorbs energy in the form of heat, its temperature can increase or it can change its state of aggregation. (You studied this in Unit 2.) An increase in temperature implies an increase in the agitation of particles. This produces changes in the properties of the object.

THINK AND REASON

If you put a dented (but not broken) ping pong ball into a pan of boiling water, the dents disappear. The ball contains the same amount of air when it is dented as when it is repaired. Why have the dents disappeared?

To understand what happens when a substance absorbs thermal energy, you must take into account what happens at both a macroscopic and molecular scale.

The object absorbs thermal energy. Therefore, the agitation of its particles increases. → As a consequence, the separation between the particles increases. → Macroscopically, the increase in agitation corresponds to an increase in temperature. The increased separation between the particles results in an increase in the volume of the object. We call this **expansion.**

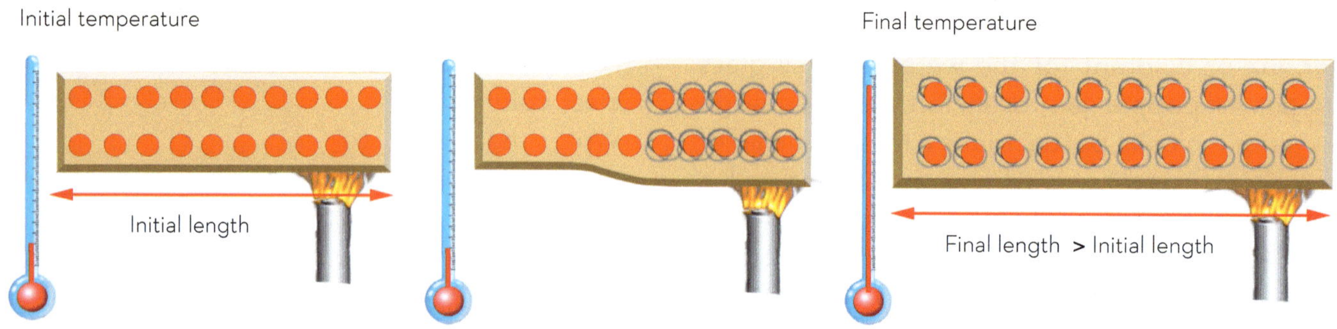

The more intensely united the particles, the less the material expands. Remember what you know about the states of aggregation. This will help you understand why gases expand more than liquids and why liquids expand more than solids.

ACTIVITIES

15. Find information about the expansion joints used in paving, bridges and buildings. Record a video talking about the importance and usefulness of these elements.

3.1. The thermometer

Liquid **thermometers** are instruments that measure the temperature of an object. They are based on the expansion of a liquid when its temperature varies.

The molecules collide in the two liquids and against the wall of the thermometer. This allows the transfer of energy between the two liquids by conduction until they reach **thermal equilibrium.**

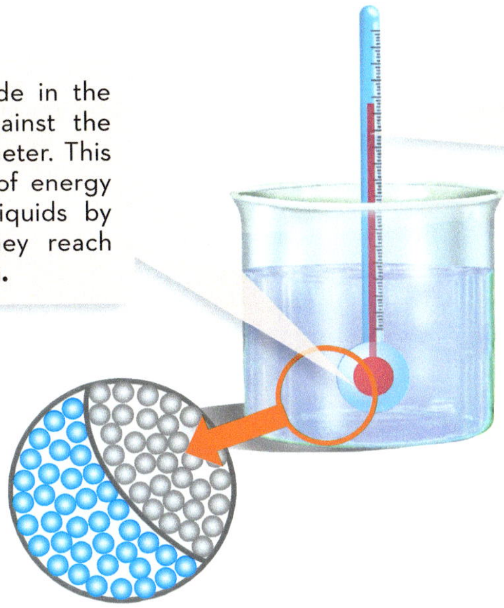

When the alcohol absorbs thermal energy, it expands and rises up the thermometer tube. The tube is graduated and indicates the temperature of the alcohol. This therefore tells us the temperature of the object.

WORKSHEET

Download this unit's worksheet and do activities 14 and 15.

REMEMBER

The equivalence between Celsius and Kelvin degrees is:

T (K) = T (°C) + 273

The **Celsius temperature scale** uses the melting point of ice and the boiling point of water as a reference. It divides the difference between these two temperatures by one hundred to define one degree Celsius (°C).

EXPERIMENT

4. You are going to make a homemade thermometer and calibrate it in degrees centigrade. You will need: water, rubbing alcohol, colorant, a plastic bottle, plasticine, a straw, a sheet of paper, a roll of adhesive tape and a dropper.

   Mix equal parts of water and alcohol. Dye it with the colorant and fill the bottle with the mixture. Construct an instrument like the one in the picture. The modelling clay must seal the bottle opening completely. Partially fill the straw with liquid using the dropper.

   Put some ice into a bowl and wait for it to partially melt. Put the bottle into the bowl. Mark the height of the liquid in the straw as 0 °C. Now do the same with a bowl containing boiling water and mark the height of the liquid in the straw as 100 °C. Now divide the distance between 0 °C and 100 °C into equal parts.

   - Can we use our thermometer to measure a temperature below 0 °C?
   - Put your thermometer on top of a heated radiator and compare your measurement with the value given by a commercial thermometer.

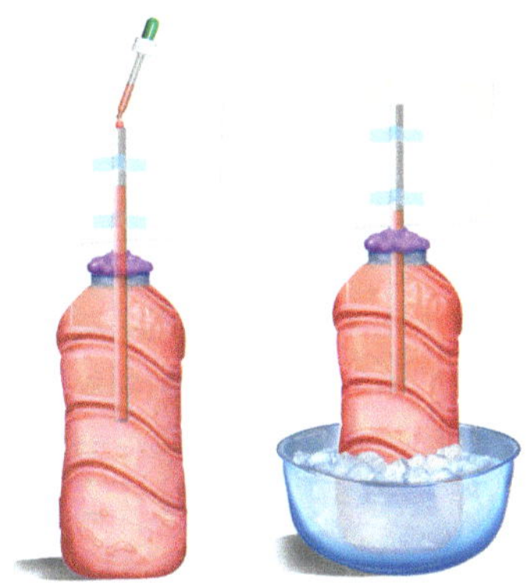

WORKSHEET
Download this unit's worksheet and do activity 16.

4 Energy in our lives

Throughout our lives we consume energy in various forms: **heat, light, motion, sound,** etc.

THINK AND REASON

The map below shows the energy consumption per person in each country of the world.

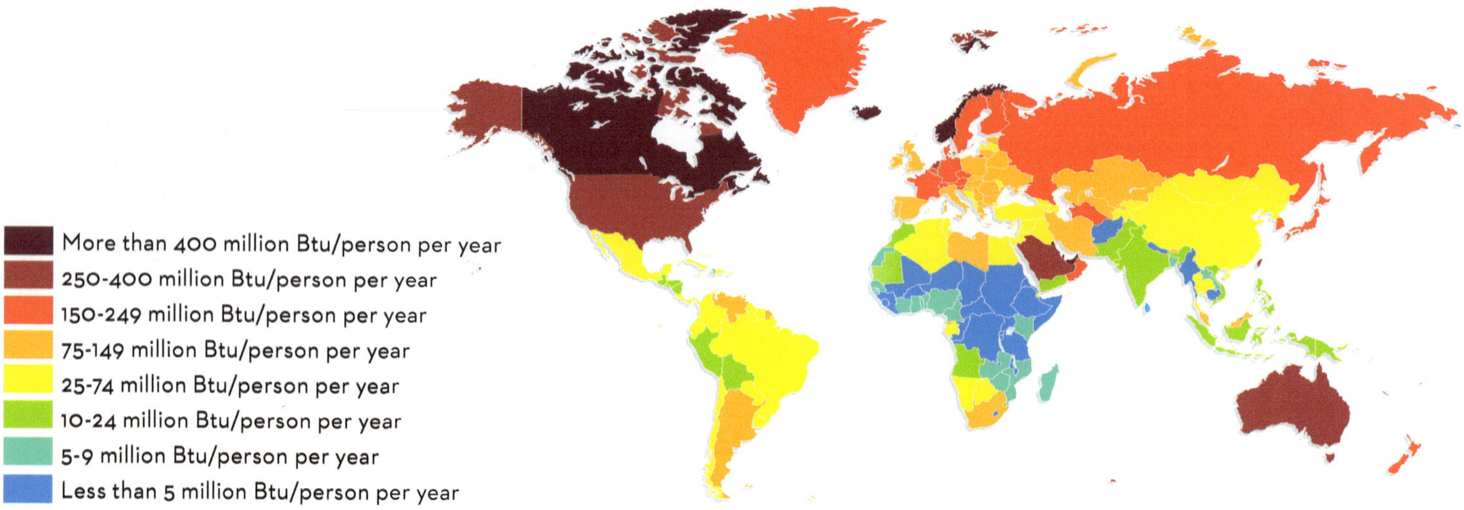

- Which countries' citizens consume the most energy per person? Is there a relationship between this consumption and the degree of the country's economical and social development?
- China is the first or second economic world power, but its energy consumption per person is relatively low. Why do you think this is?

DID YOU KNOW?

Approximately 25% of the energy generated in the world each year dissipates as it is generated or during its transportation. Therefore, 25% of the energy generated is useless energy.

Humans obtain the energy we consume from different resources we call sources of energy.

An **energy source** is any resource from which humans extract energy that transforms in accordance with their needs. Energy sources can be **renewable** or **non-renewable**.

Nowadays, the energy obtained from these sources has three uses:

- We mainly use non-renewable energy sources for **transport.**
- We mainly use non-renewable energy sources and a smaller proportion of renewable sources to produce **heat** for the **central heating** in our homes and **industrial facilities.**
- We produce electrical energy for **electrical devices** with renewable and non-renewable sources.

4.1. Non-renewable energy sources

Non-renewable energy sources are finite. This means that we consume them at a faster rate than they are created, so they eventually run out.

OIL

We extract oil from underground wells. We use it to obtain different combustible substances (fuels such as petrol, diesel and kerosene), but also plastics or asphalt.

COAL

We extract coal from underground mines. We mostly use it to obtain electrical energy in thermal power stations.

NATURAL GAS

We extract gas from underground wells. We use it for cooking, central heating and industrial furnaces. We also use it to obtain electrical energy in thermal power stations.

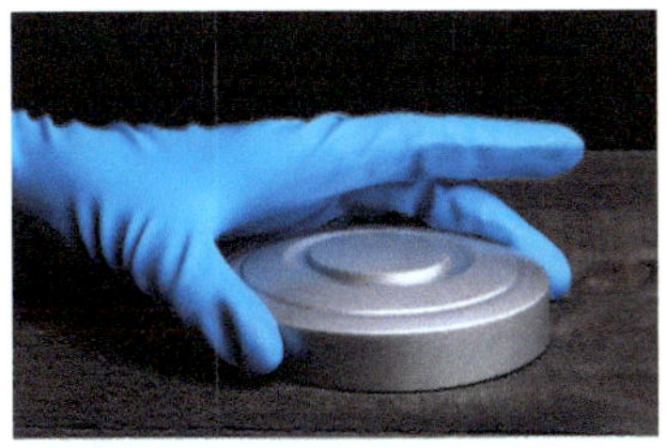

URANIUM

We obtain uranium from underground mines. It has to be concentrated. It is the main fuel in nuclear power stations. These use nuclear fission to obtain electrical energy.

We use the term **fossil fuels** to refer to **oil**, **coal** and **natural gas**. Fossil fuels and **uranium** are the most widely used non-renewable energy sources at present.

Oil is the main energy source, as it allows humans to obtain the fuels needed to transport people or goods.

WORKSHEET

Download this unit's worksheet and do activities 17 and 18.

ACTIVITIES

16. The following diagram shows the origin of the energy consumed in the world in 2006. What percentage of the energy used is from a non-renewable source?

17. State three examples from your everyday life where you consume energy for each one of the three main uses described above. Share them with the rest of the class and make a list in your notebook with all the examples.

18. Investigate the controversy that arose in the Canary Islands about the possibility of surveying for oil deposits there. Prepare a brief oral presentation in favour of these operations and another one against them and show them to your class.

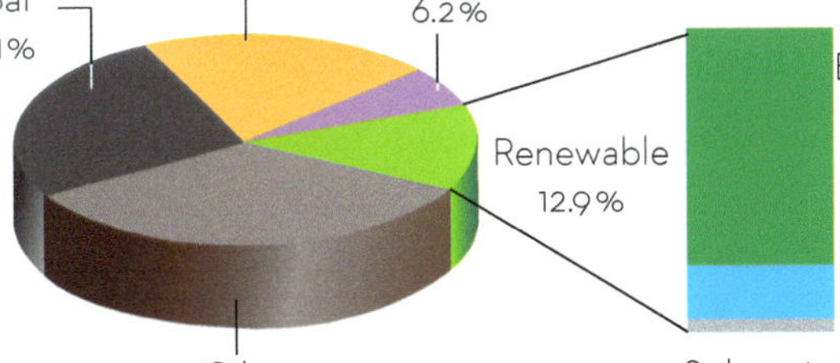

4.2. Renewable energy sources

Renewable energy sources are inexhaustible. This means that we consume them at a lower rate than nature creates them, so they never run out.

SOLAR PHOTOVOLTAIC

Solar panels convert sunlight into an electric current. This process can only take place during daylight hours and this is one of its major problems.

SOLAR THERMAL

Sunlight heats the water circulating through pipes. We can use this hot water directly in the kitchen and bathroom or for central heating.

GEOTHERMAL

The Earth's internal heat transforms liquid water into steam. The steam propels turbines, which transform the motion into electric current.

TIDAL

The tides, waves and currents in the sea move the blades of underground turbines. The turbines then transform this motion into electric current.

BIOMASS RESIDUES

We can transform the waste from crops, cattle or water treatment into fuel for central heating or industrial furnaces.

BIOFUELS

We can convert crops such as sunflowers, soya or sugar cane into fuels like bioethanol or biodiesel. They substitute fossil fuels.

LAND OR SEA WIND

Wind moves the blades of the turbines, which transform this motion into electrical energy.

HYDROELECTRICAL

The water in dams flows through pipes with turbines at the end. The turbines transform this motion into electrical energy.

ACTIVITIES

19. Choose one of the eight types of renewable energy and create a presentation about it. The presentation can be: a slide show, a poster, a documentary video, an audio production (similar to a radio show), a blog, etc.

4.3. Renewable or non-renewable energy?

Access to energy is fundamental to the development of societies. Over the next few pages, we are going to compare different aspects of renewable and non-renewable energy sources. This will help you to reflect on how humans should manage the way we obtain energy.

WORKSHEET

Download this unit's worksheet and do activities 19 and 20.

A. Geographical and economic aspects

One of the main objectives of every country in the world is to have a large enough energy supply. They need energy for their industries and services, and for their citizens to have an adequate standard of living.

THINK AND REASON

Observe the following data from 2014 on the main oil-producing countries and the countries that produce electricity from renewable sources:

Main producers of oil (percentage of the total number of barrels in the world produced by each country)

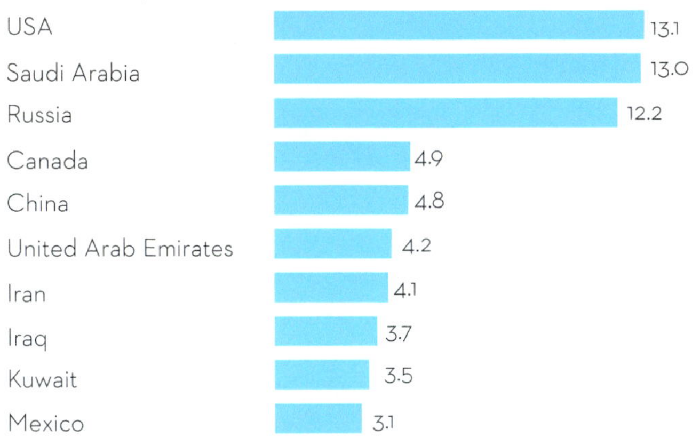

Source: BP

Main producers of electrical energy from renewable sources

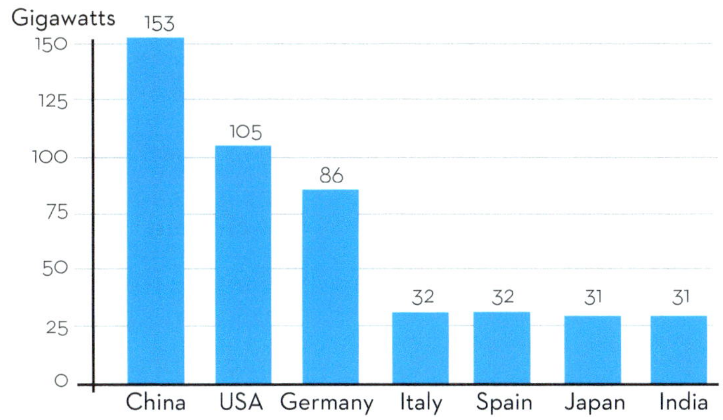

Source: REN 21

- No European country is an important oil producer, but three of them (including Spain) are large producers of electricity from renewable sources. Do you think there is a relationship between these two facts?
- Why do you think that China and the United States are investing in renewable energies, even though they produce lots of oil? Why aren't countries such as Iran, Iraq and Kuwait producing renewable energies?

Renewable energies are not limited to specific territories. They can be used by all countries.

Renewable energies allow many countries to overcome their **foreign energy dependence**. This dependence means they have to buy fuel from other countries and are therefore partially dependent on them.

There is a lot of economic power associated with energy. Large companies and individual countries try to obtain the best position and ensure they have sufficient energy resources.

B. Environmental aspects

Any method of obtaining energy produces a change in the environment. In particular, non-renewable energies cause serious problems with grave consequences for mankind and nature.

Climate change

This is the greatest environmental problem humans face at present. Burning fossil fuels emits **carbon dioxide (CO_2)**, which accumulates in the atmosphere. This gas absorbs and then emits again part of the solar radiation which should leave the planet. This causes the temperature to rise, leading to droughts, torrential rains, tornados, polar ice melt, alterations in the lifecycle of plants and animals, etc.

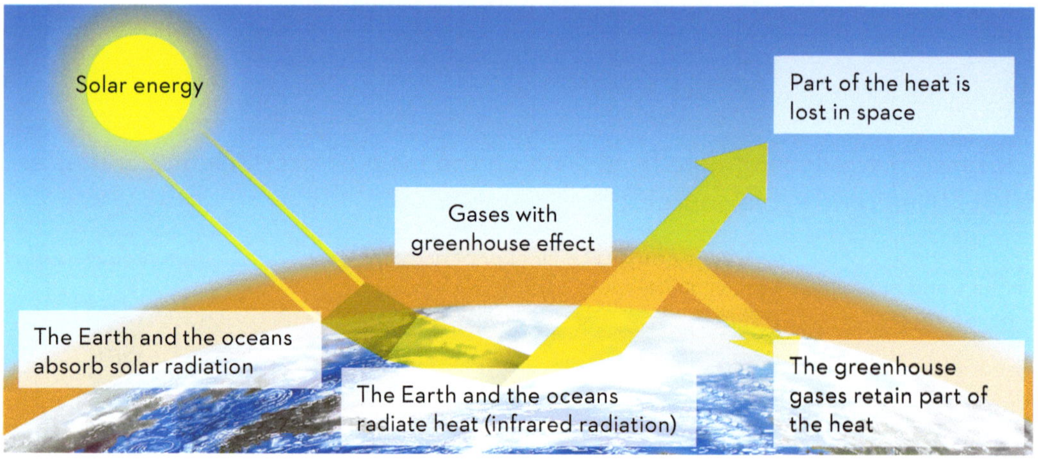

Acid rain

Burning fossil fuels emits nitrogen oxide and **sulphur** dioxide into the atmosphere. When these oxides come into contact with the water in clouds, they transform into sulphuric and **nitric** acids. These acids return to the ground dissolved in rainwater. This causes serious problems in lakes, soils and stone buildings.

Radioactive waste and nuclear accidents

Nuclear power stations produce highly carcinogenic and toxic radioactive waste. This waste can be active for thousands of years. In addition, the power stations can have accidents if their reactors are out of control. This has very serious consequences for humans and the environment.

DID YOU KNOW?

The ITER project is an international project to obtain energy through **nuclear fusion**. This energy source would be almost inexhaustible and would not generate environmental problems. There is currently intense research into nuclear fusion. Scientists are slowly resolving the enormous technical problems it presents. However, they are still not ready to confirm that it will become a stable new source of energy for mankind.

WORKSHEET

Download this unit's worksheet and do activities 21 and 22.

C. Technological aspects

Non-renewable energies use consolidated technology. Petrol or diesel engines and electricity-generating thermal or nuclear power stations are perfectly linked to our societies. Renewable energies must still complete this adaptation.

THINK AND REASON

The graph shows electricity consumption in Spain over a single day.

- Describe the graph and interpret the peaks and troughs in electricity consumption, thinking about the Spanish lifestyle.
- What is the problem with renewable energies to meet the peak hours of demand? Does the wind blow with the same force all the time?

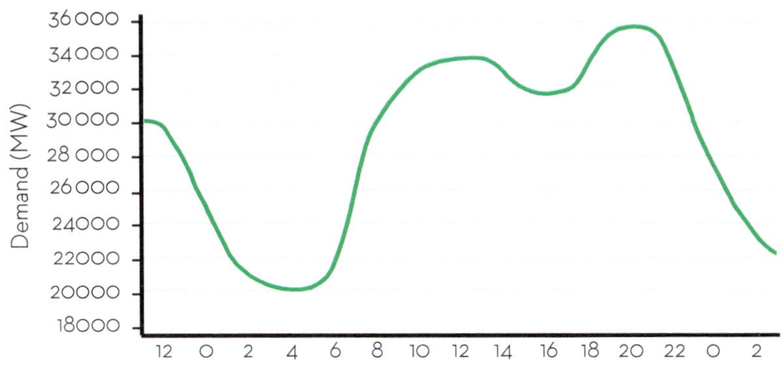

Source: REE

Renewable energies currently provide around 51% of the electricity consumed in Spain. To increase this percentage, we need to implement two measures:

- **Improve energy storage.** We need to make energy from renewable sources available when the wind is not blowing or there is no sunlight.
- **Modify the consumption habits** of the population. We must commit to using energy at the most propitious times for renewable energies.

As well as obtaining electrical energy from renewable sources, the other great change to our energy consumption will come from the use of **electric cars**.

There are currently vehicles which are completely electric and there are others that are **hybrids**. These have two engines: one is for fossil fuel and the other is electric.

These vehicles charge their batteries by plugging them into an electric current with a normal socket or a fast-charge socket. The main problem is still their battery range, but scientists are working hard to improve this. At the moment they need to be recharged approximately every 200 km. As a result, they are not very useful for long journeys.

ENERGY 2.0

Go to the following link:

goo.gl/hUwxjp

You can see the origin of the electrical energy consumed in Spain hour by hour for any day of the year. Work with your partner to locate in which month of the year we consume the most and try to explain it.

ACTIVITIES

20. Renewable energies present far fewer problems for the environment than non-renewable ones. Investigate and describe the environmental problems presented by solar, wind, hydroelectric and biofuel energies.

21. Look at the following link: **goo.gl/hUwxjp** Compare the origin of the electrical energy consumed in Spain during the day and during the night. Now compare the electrical energy consumed in a winter day and in a summer day.

5 The need for energy saving

WORKSHEET

Download this unit's worksheet and do activities 23 and 24.

Mankind is consuming increasing amounts of energy. Obtaining this energy is causing serious problems and this is forcing us to consider energy saving as a priority for our societies.

THINK AND REASON

Observe the following figures:

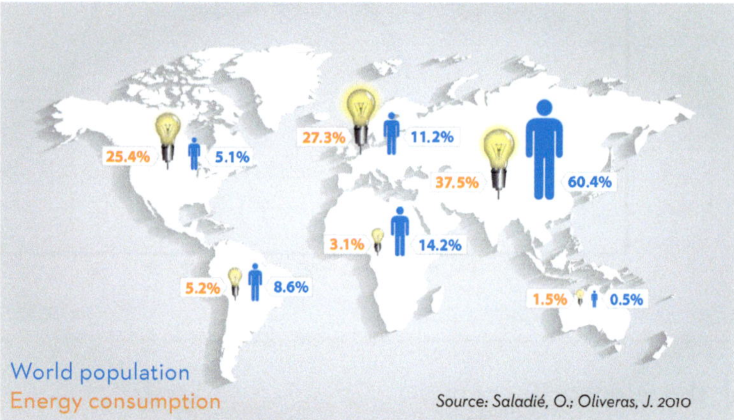

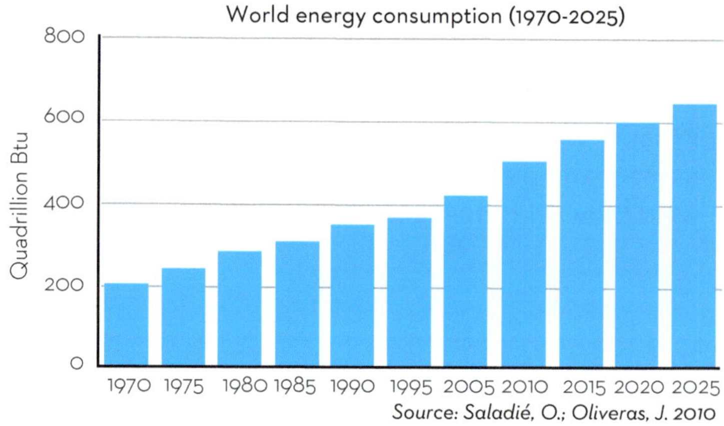

Based on the data on the map, what differences do you think there are between a North American person's life and that of an African person?

The world's population is increasing. Do you think world energy consumption will continue to increase? Why? Who has a greater responsibility to save energy: a European country or a South American one?

The current growth in world energy consumption is unsustainable. The main reasons are as follows:

Climate change is a real threat to the future of mankind. Scientists believe we have to reduce the burning of fossil fuels dramatically. Consequently, we have to reduce the energy obtained in this way.

Fossil fuels are **finite**. If we consume them at our current rate, they will run out in the medium term. In addition, these resources are very useful for obtaining plastics, for example. Some experts consider it a waste to burn them as fuel.

Renewable energies are an excellent energy source, but they cannot meet the growing **energy demand.** There is currently a lot of research into increasing its efficiency and finding new energy sources that complement them.

5.1. How can we save energy?

There are many things we can do in our everyday lives to save energy.

AT HOME

- Always use electrical appliances when they are full and buy the most efficient ones. Look at the diagram explaining the meaning of each energy class so you can interpret the meaning of the labels on your appliances.
- Recycle everything you can.
- Cook with a pressure cooker. Don't pre-heat the oven if it isn't necessary. Put the lid on pans when cooking on the hob.
- Use low-energy lighting and only turn on the lights when you need them.
- Don't leave devices plugged in when they are switched off or on standby.
- In winter, put the heating on between 19 and 21 ºC. Put the air conditioning on at 7 or 8 ºC below the external temperature.

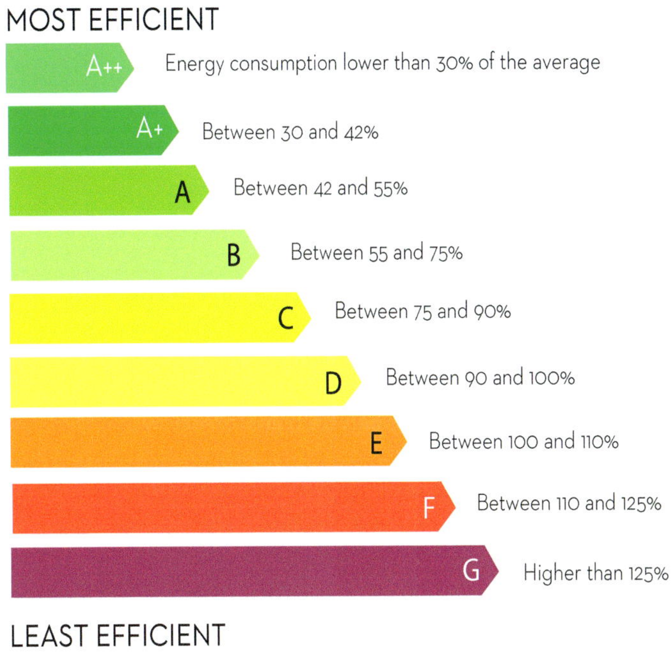

TRANSPORT

Walk or ride a bike for short journeys. Share the car with other passengers. Use public transport.

SHOPPING

Try to buy locally produced food that is in season. When buying clothes and technology, only buy goods you really need.

IN YOUR CITY, AT WORK, AT SCHOOL

When you see bad energy habits, always tell someone who can change them. For example, the mayor in your town, your headteacher, etc.

ACTIVITIES

22. Find information on the internet about world energy consumption from the mid-19th century until the present. How much has this consumption multiplied? Give several reasons for this increase. Compare your conclusions with your partner.

23. Create a list of ten energy-saving measures to apply in your school. Compare all the lists in the class and create a new final list of the ten best measures. Make a presentation with the items in the list and show it to the class.

MIND MAP

Download and print the worksheet you will find in the OLC to complete the mind map of the unit.

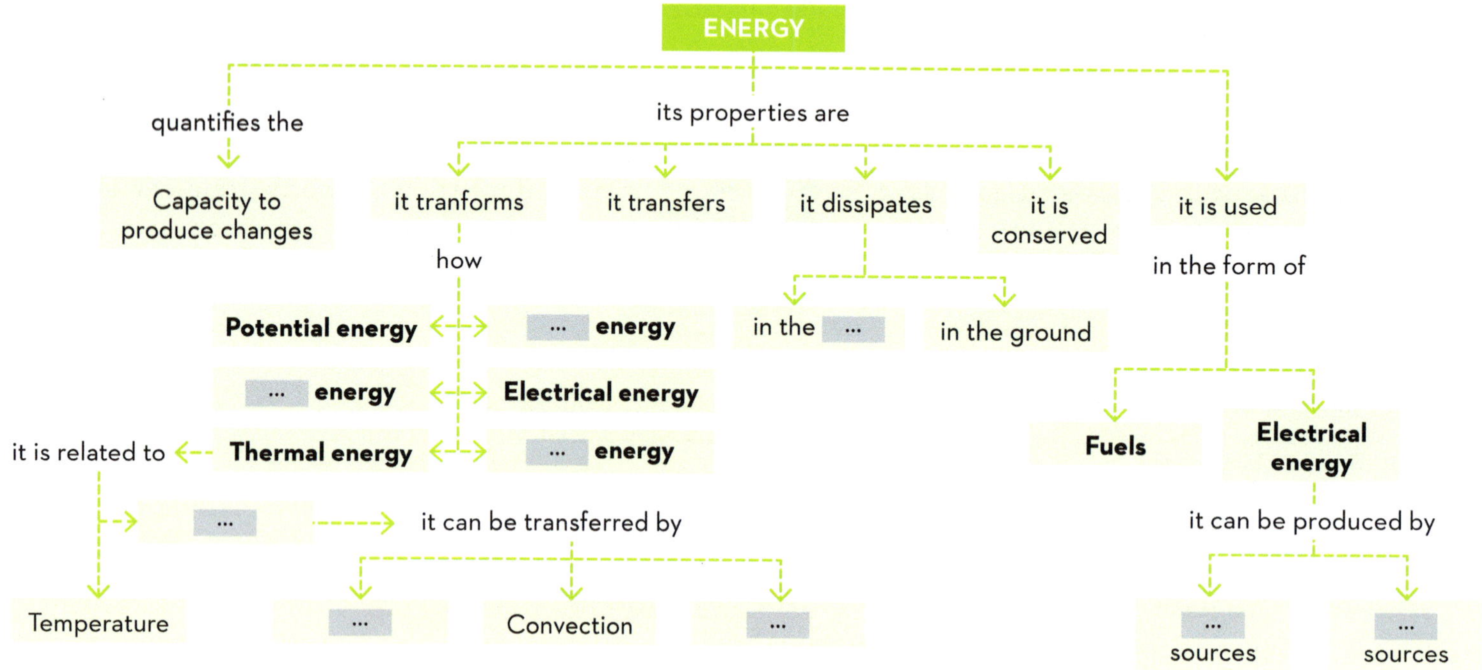

LOOK AROUND YOU

Wind provides El Hierro with all its electricity for four hours

The island of El Hierro reached a world record. It is the first isolated territory in the world to achieve 100% of its energy supply from a renewable energy source. At 12 o'clock midday on Sunday 9th August, the Gorona del Viento hydro-wind power station began to generate the entire electricity supply for the island from clean sources. It maintained the supply for four hours.

The technicians achieved this thanks to the combination of two sources of renewable energy: hydraulic power from the Gorona power station coupled with a wind farm with five wind turbines. This system will save 80 million euros over the next two decades, because 6000 fewer tonnes of diesel will be burned each year. The project cost 82 million euros and represents an innovative technological achievement.

Source: El País. 20th August 2015.

Questions

a) Where is the island of El Hierro? What is the size of its population? Does it have any important meteorological characteristics?

b) What positive environmental aspects does a completely renewable energy supply have? Why do experts believe that different renewable energies should always be combined to work together?

c) Do you think that your city could be supplied only by renewable energy sources? Discuss your answer with your partner and make a proposal for your city's mayor on how your city can be supplied only by renewable sources.

LABORATORY WORK

Energy, thermal conductivity and expansion

Objective

The objectives of this experimental class are to: study the transformation of energy from one kind to another; compare the conductive capacity of different metals; and experimentally observe the process of thermal expansion.

Material

Glass jar with lid (a jam jar, for example), sand, thermal conductivity apparatus, Bunsen burner, wax or paraffin, and expansion ball and ring apparatus.

Thermal conductivity measurement apparatus

Methodology

a) Fill the glass jar with sand to approximately one third full. Measure the temperature of the sand in the jar. Close the jar and shake the sand hard for one minute. If you get tired, pass the jar to one of your classmates to continue shaking it. After one minute, open the jar quickly and measure the temperature of the sand again.

b) Use the thermal conductivity measurement apparatus. (Remember, thermal conductivity is the capacity of different materials to conduct thermal energy.) Put a portion of wax or paraffin on the end of each metal wand. Heat the centre of the apparatus with the Bunsen burner. Observe the order in which each portion of wax melts.

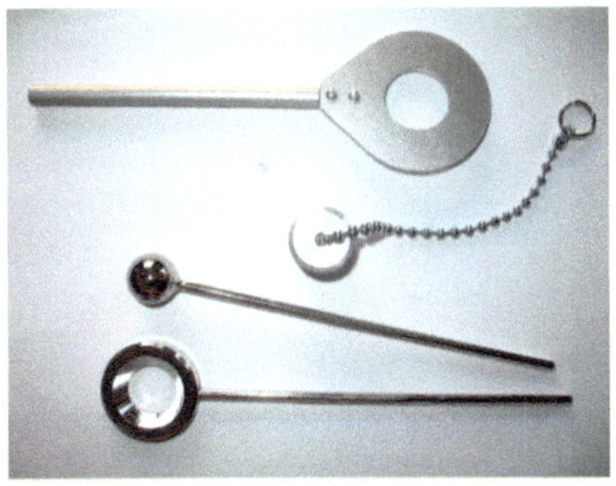

Metal ball and ring apparatus

c) Finally, we are going to monitor the thermal expansion of a solid. To do this, we are going to use a ball and ring apparatus. First, check that the ball can pass through the ring. Now, with the ball inside the ring, heat the ball with the Bunsen burner for a few minutes. Then try to take the ball out of the ring.

Questions

- In experiment *a)*, what energy transformation has taken place? The grains of sand have increased in temperature. What does that indicate about their properties?
- In experiment *b)*, what heat transference mechanism takes place? Put the different materials of the apparatus in order, from best to worst thermal conductor and check your results with your partner. What does it mean at a particle level that a material is a good heat conductor?
- In experiment *c)*, why can't you take the ball out after you have heated it? One pupil explains it in the following way: 'When we heat the ball, its particles increase in size. As a result, the volume of the ball increases and cannot be removed from the ring.' Is this explanation correct? Can you clarify what really happens? Discuss this with your partner.

FINAL ACTIVITIES

Basic activities

1. Design an A3-sized poem poster which includes several images of systems with changes happening in them. Explain the types of energy there are. The title of the poster is: 'Where there is change, there is energy.' It should contain a poem about energy and its transformations. On the back of the poster, write a list of the sources you have used to make the poster.

2. There is one receptacle containing three litres of water at 20 ºC on a hot plate and another receptacle containing one litre of water at 20 ºC on an identical hot plate. Twenty minutes later:

a) Will the same amount of energy be transferred to both receptacles?

b) Will the water in the two receptacles be the same temperature? Which one will be higher?

c) Will the molecules of water in both receptacles have the same kinetic energy? Which one will have more?

d) Can you compare the above situation to a father sharing 100 euros with his two children and another father sharing 100 euros with his five children? Explain the analogy.

Take turns with your partner to answer the questions and reach an agreement.

3. A car travelling at 80 km/h possesses a large amount of kinetic energy. When the driver steps on the brakes and the car stops, what is this energy transformed into? Is it transferred to another object or does it dissipate? Is the principle of energy conservation true here? Why? Compare your answers with your partner.

4. We have all burned ourselves eating croquettes at some point. When we pick them up with our fingers, they don't feel very hot. However, when we bite into them, we burn our mouth. Does the coating on the croquette have a thermal conductivity similar to metal or to plastic?

5. The iron in the picture has reached a temperature of 120 ºC. Working in pairs, answer the following questions: How will the temperature on the thermometers change after coming into contact with the iron for a short time? What do we mean by: 'thermal equilibrium will be reached' by maintaining contact?

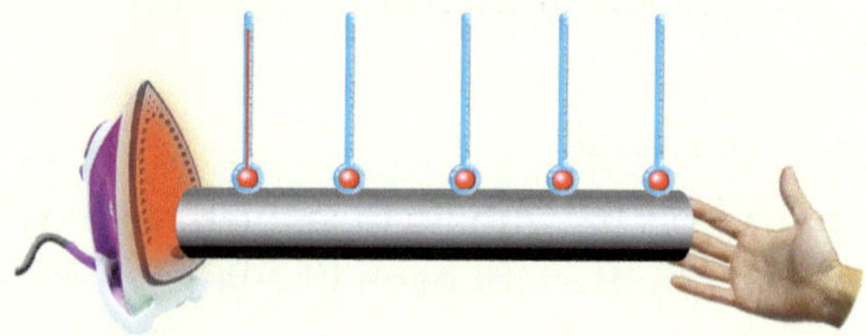

6. A 100 gram piece of iron at 90 ºC is put into a receptacle containing a litre of water at 25 ºC. Which way does the transfer of heat go? Describe how this energy transference occurs.

7. The image shows a simplified explanation of the movement of tectonic plates. Explain the diagram and indicate the implied heat transfer mechanism.

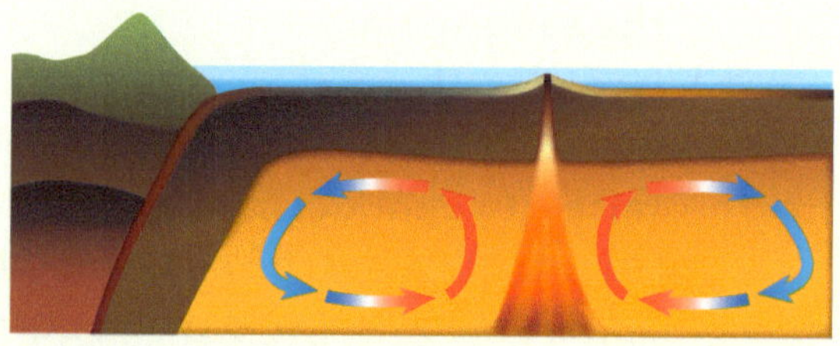

8. Why does installing an awning over windows and terraces help to reduce the temperature inside the house? Which thermal energy transfer mechanism is important here? Discuss this with your partner.

9. A mechanic wants to join a bar and a ring together. He cannot remember if he has to adjust the hot ring inside the cold bar or the cold ring inside the hot bar. Can you help him?

10. With the help of your partner, search for information on the current situation and future prospects for energy from nuclear fission. Are there more or fewer nuclear power stations in the world now? Why do you think this is?

11. In December 2015 there was a Climate Summit to find solutions to the problem of climate change. Prepare an elevator pitch to convince the heads of government of the need to change our energy sources and our current mode of consumption.

Consolidation activities

12. Explain the energy transformation that occurs when somebody jumps down a slide at a water park. How is energy being transferred or dissipated in this situation?

13. Record a video explaining energy conservation based on the functioning of a light bulb. Pay attention to three aspects: the electricity supply, the emission of light and the temperature the bulb reaches when it is on.

14. Work in pairs. Draw a bar chart showing the division between electromagnetic (light), sound and thermal energy in a mobile. Estimate the percentage of each type of energy as if they were proportional to the time the mobile emits light or sound and dissipates energy in the form of heat.

15. We have three different liquids to make a thermometer. Liquid A triples in volume when it goes from 15 to 50 ºC. Liquid B multiplies its volume by 1.2 in the same process. Liquid C multiplies its volume by 6. Which is the best liquid to make the thermometer? What is the problem with the other two? Think how long the thermometer would have to be if you used those liquids.

16. Observe the data on energy consumption in the home. Where do you think it is most feasible to save energy? Suggest two energy-saving measures for each section on the graph. Share them with the class and make a poster with the best proposals.

Advanced activities

17. Which consumes more energy: a television with a 10-inch screen or one with a 40-inch screen?

18. David takes an orange out of the fridge. He squeezes it and puts the glass of juice in the microwave for thirty seconds. Andrew puts the whole orange in the microwave for the same amount of time and then squeezes it.

Which receives more thermal energy: the glass of juice or the whole orange? Which glass of juice will have a higher temperature? Explain your answer using the principle of energy conservation.

19. Think, pair up and share: indicate three situations in your everyday life where the energy dissipation is positive and three situations where it is negative. Share your situations with your partner and then with the whole class.

20. Some isothermal bottles consist of a container surrounded by a vacuum chamber. In other words, there is an absence of matter inside the chamber. Explain why this system keeps the liquid inside the flask at a constant temperature.

21. At the beginning of December 2013, there was very little wind in Spain. A few days later, an explosive cyclogenesis produced an enormous amount of wind. What does this information tell you about the limitations of wind energy? We define drought as the absence of rain for a long period of time. What energy source does drought affect?

22. Reflect on the importance of being able to store electrical energy at a competitive price to improve the implementation of renewable energies. Write a newspaper article. **NOTE:** You can download the parts of an article from the OLC.

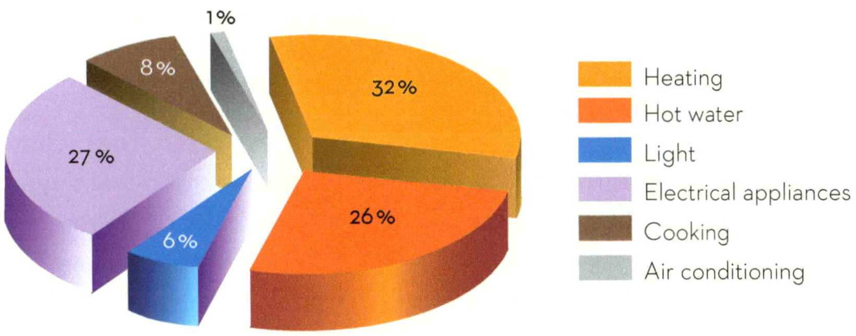

The greenhouse effect: reality or fiction?

It is a fact that the average temperature of the atmosphere has increased. People often say that the main cause of the rise in temperature in the 20th century is carbon dioxide emissions.

A student called Andrew is interested in the possible relationship between the average temperature of the Earth's atmosphere and the carbon dioxide emissions on Earth. He finds the following graph in the library.

Based on this graph, Andrew concludes that it is true that the rise in the average temperature of the Earth's atmosphere is caused by the increase in carbon dioxide emissions.

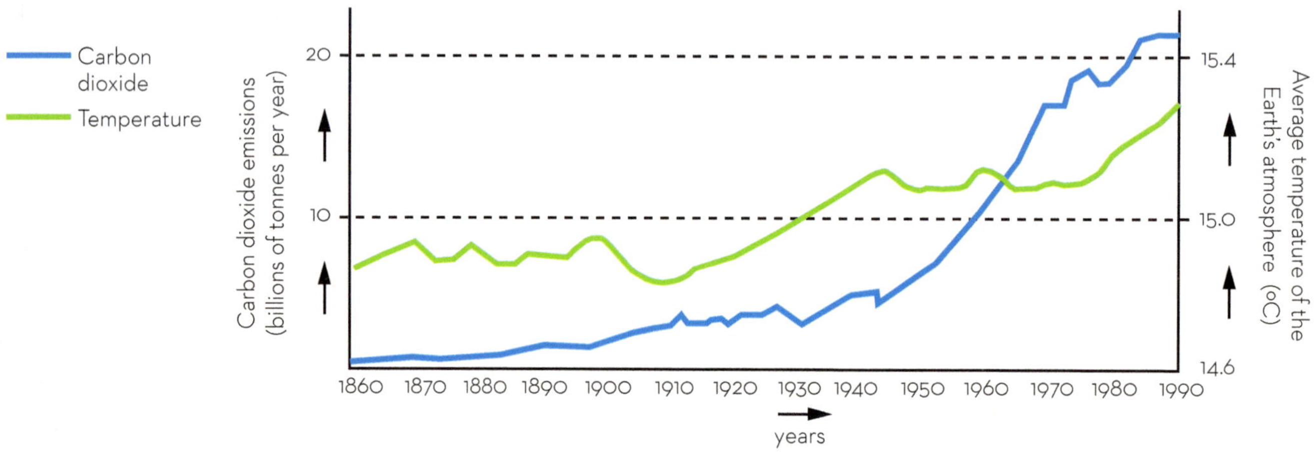

Working in pairs, answer the following questions about Andrew's work:

Question 1

What information in the graph supports Andrew's conclusion?

Question 2

Another pupil, Joanne, does not agree with Andrew's conclusion. She compares the two lines and says that some parts of the graph do not support his conclusion.

Select one part of the graph which does not confirm Andrew's conclusion. Explain your answer.

Question 3

Andrew insists that the cause for the rise in the average temperature of the Earth's atmosphere is the increase in carbon dioxide emissions. But Joanne thinks his conclusion is premature. She says, 'Before accepting this conclusion, you must check that the other factors influencing the greenhouse effect have remained constant.'

Name one of the factors Joanne is referring to.

Question 4

Andrew wants to start a debate about global warming at his school. He wants to invite six participants to the debate. Which six professions do you think would be best to explain the entire phenomenon? Follow the example: A biologist can talk to us about how climate change is affecting animals.

Question 5

Andrew reads that 'climate change' is the preferred term nowadays, not 'global warming'. What do both terms mean? Why do you think it is preferable to use 'climate change'? Use the following structure to answer the questions: Global warming means...

Mobilise the world and save energy

Competency task

Objective

The objective of this task is to create an energy-saving campaign involving your friends, classmates, families, etc. You can do this task individually, but it will be much more effective if the whole class does it as a team.

Final product

The final product you must create is a sign motivating people to join the energy-saving campaign. You must also think of a way to count the people who join your campaign.

Step by step

1. Make a list of five energy-saving goals you are going to propose to various people you know. These five goals must be feasible, be effective and have varying degrees of difficulty. For example, a simple goal might be: 'I turn off all the lights I don't need. I don't waste a single photon of light.'
2. Assign a medal to each goal. For example, wood, plastic, bronze, silver and gold medals (from the simplest to the most complex goal).
3. Create an attractive logo for each goal. Draw the five medals with their respective logos in the centre. For instance, you could set the example goal above to earn the plastic medal. The diagram shows a possible logo.
4. Create a sign showing the five medals with the five goals next to them. If the whole class is doing the project together, you can hang the sign in a prominent place in your school. In that way, all the pupils in the school can join the campaign.
5. Make an individual medal counter or one for the whole class. If it is for the whole class, make a poster and put it up in the classroom.
6. Use all your resources. First, choose the goal you are going to achieve, photograph the corresponding medal and make stickers, badges or something else with this photo. Share these emblems among people you know who promise to reach the goal.
7. Ask your teachers if you can go to other classes in the school to explain the project.
8. Every time somebody (pupil, friend, family member, etc.) promises to achieve one of the goals, add them to your medal counter.
9. Tell everyone involved how many medals you have earned. Encourage everyone you know to reach a common goal. 250 medals? 500 medals? 1000 medals?

With every medal you earn, you are doing the Earth a favour. Earn as many as you can!

Self-assessment

Evaluate your work and your classmates' work with the guidelines that you can download from the OLC.

8 ELECTRIC CURRENT

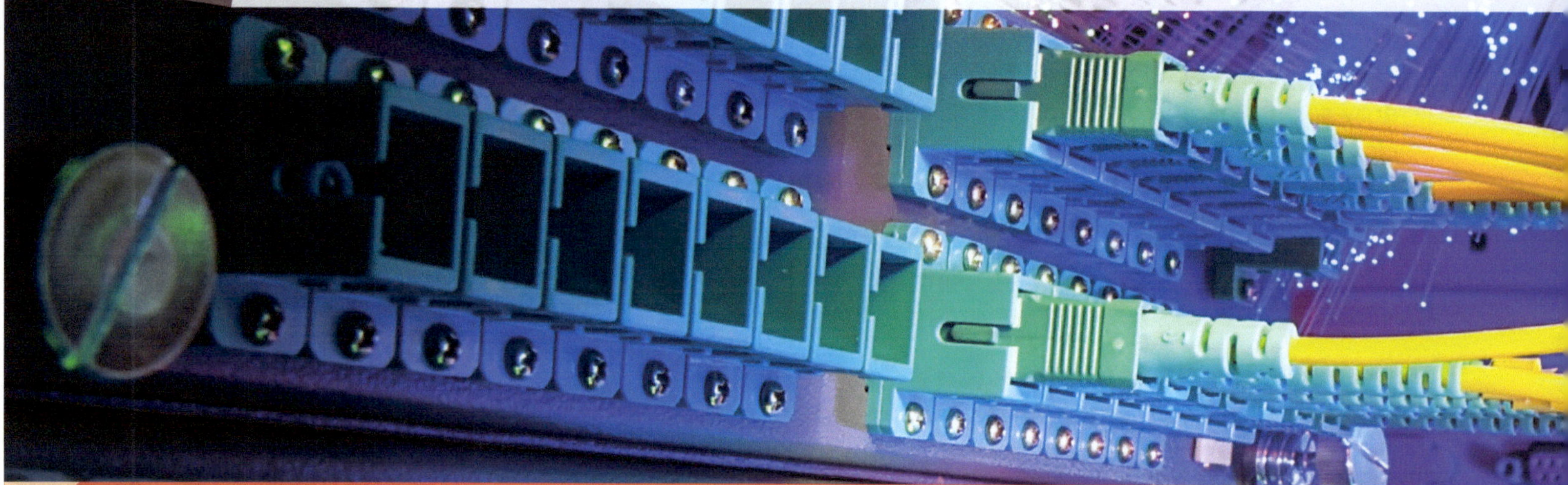

Our world today seems very complicated, but we can try to understand it better. Most changes that happen around you occur thanks to muscle power, combustions (like the movement of all vehicles) or electricity. Electricity is omnipresent in your life: mobile telephones, microwave ovens, televisions, computers and the electrical devices that control industrial processes.

An electric current consists of many electrons moving all together in the same direction along a conductor material, such as a cable of copper. If we know how to make these electrons move as we want, we will understand the functioning of an enormous quantity of modern devices and a large part of our present-day world.

To understand other more complex devices like mobile telephones or computers, we should learn how to use electrons to transport information. This is the field of work of electronics. Electronics is a science that was born in the early 20th century. It is the discipline that has caused one of the largest revolutions of all the history of humanity: the current information technology revolution.

In regard to the importance of information technology, there is an ingenious saying that illustrates it perfectly:

> 'There are only 10 types of people: those who understand binary and those who don't.'
>
> Which of the two types of people are you?

WORK WITH THE TEXT

Answer the following questions about the text:

a) What is an electric current?

b) What science studies how to use electrons to transport information?

c) Name five electronic devices you have at home.

d) Did you understand the saying at the end of the reading? Search for information on the internet about the binary system and try to explain it to your classmates.

CHALLENGE

You probably know how to use an electrical device, but are you able to produce an electric current?

To do this, you will need eight five-cent euro coins, card, vinegar, aluminium foil, copper cable, a cup, electrical tape and an LED.

Cut some square pieces of card and put them in a cup of vinegar for ten minutes. Also cut some square pieces of aluminium foil.

After the ten minutes, build a tower by putting all the elements in the following order: aluminium foil, square of card, coin. Do this successively with all eight coins. You should finish the tower with a coin. Now connect the LED to the tower of coins (the negative pole of the LED to a coin and the postive to a coin). What happens to the LED?

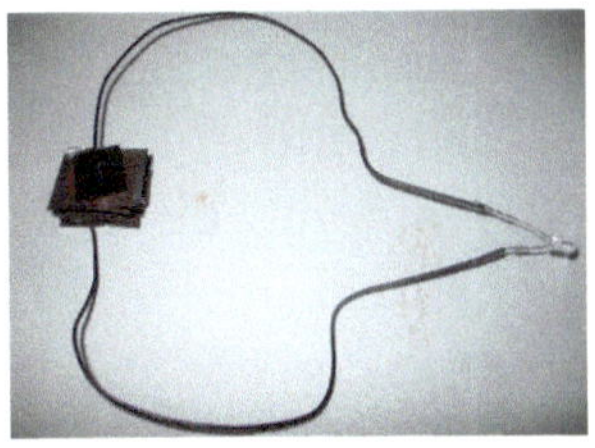

Contents

KEYWORDS

Download and print the activities that will guide you through the essential vocabulary of this unit.

FIND A CLASSMATE

Download and print the worksheet 'Find a classmate: electricity and electronics'. Share experiences about this topic with your classmates and discover things about them.

1 Electric current

The use of electric current has become indispensable in our lives. But do you know what an electric current is?

CLASSROOM LABORATORY

1. Build the circuit like in the figure with a 4.5 V battery, a lamp holder, a light bulb and a switch. Check that the light bulb lights up when you turn the switch on.
   - What type of energy does a battery have? What two types of energy does the light bulb produce? What is the intermediary between the battery energy and the light bulb energy?
   - Which element should you use to replace the light bulb if you want the battery energy to transform into kinetic energy?

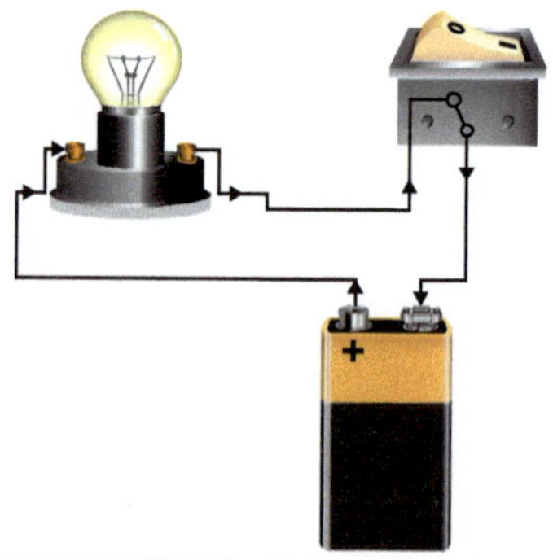

ELECTRICITY 2.0

PhET offers us a simple circuit simulator. Go to

goo.gl/zonLTb

Use it to build the circuit in the figure and observe the movement of the electrons that make up the electric current.

An **electric current** is a flow of particles with an electrical charge that move between two points through a conducting substance.

The most common example of electric current is a flow of electrons that move through a metal conductor, but there are more examples of charged particles that can make an electric current. The electrical charge is measured in coulombs (C). The charge of an electron is equal to $1.6 \cdot 10^{-19}$ C.

The great usefulness of electric current is that its energy can be transformed into all types of energy:

WORKSHEET

Download this unit's worksheet and do activities 1 and 2.

ACTIVITIES

1. Work in pairs. If we compare an electric current to a race, what do the runners correspond to? What do the streets they run in represent?

2 Electrical magnitudes

We use three magnitudes to talk about an electric current. These are the **current intensity, voltage** and **electrical resistance.**

THINK AND REASON

The figure shows a device made up of two tanks with water, a pipe connecting them and a wheel that returns the water from the lower tank to the upper tank.

How would the amount of water that falls per second to the lower tank vary if the level of water increases in the upper tank? What if the diameter of the pipe is doubled? What if the pipe is partially obstructed?

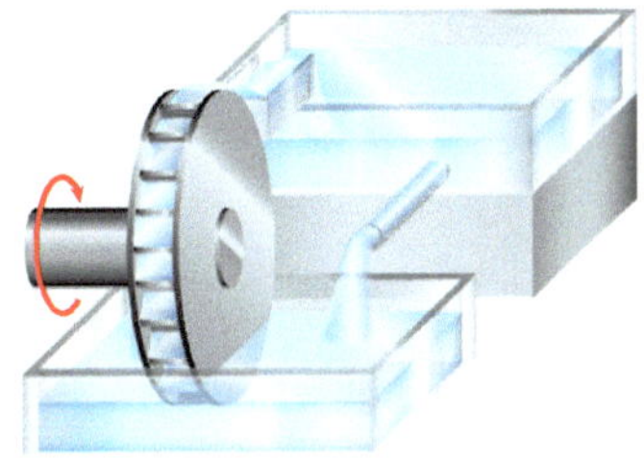

To understand the three main electrical magnitudes, it is very useful to use the analogy between electric current and water systems.

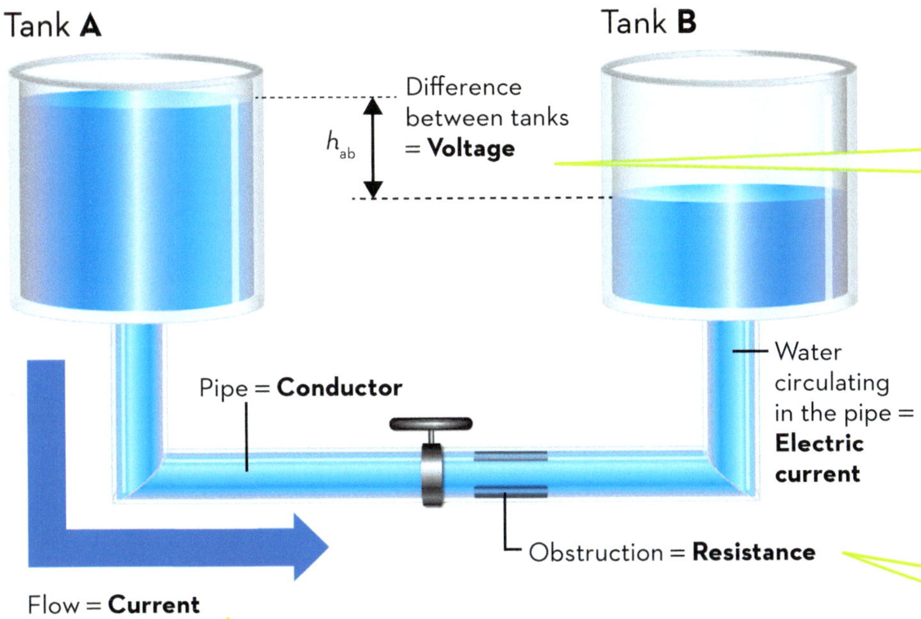

The **potential difference** or **voltage** between two points of a circuit is the impulse that the current receives to go from one point of the circuit to the other. It is measured in volts (V) in the International System. One volt is equal to one joule (unit of energy) for every coulomb (unit of electrical charge). Therefore, you can also interpret the voltage between two points as the difference of energy of the electrons between those two points.

Electrical resistance quantifies the opposition of the conductive material to the electrons circulating. Electrical resistance is measured in ohms (Ω) in the International System.

Current intensity measures the amount of electrical charge that passes through a section of conductor per second. Currents are measured in amperes (A) in the International System. A current of one ampere means that six trillion electrons pass through the conductor every second.

ACTIVITIES

2. **Continuing with the analogy in activity 1, what do current intensity, electrical resistance and the voltage between two points of a conductor correspond to in a race?**

3. **Work with your partner. Think of another analogy to explain what an electric current is and describe its three main magnitudes.**

2.1. Ohm's law

The three electrical magnitudes described on the previous page are related by a very simple expression.

THINK AND REASON

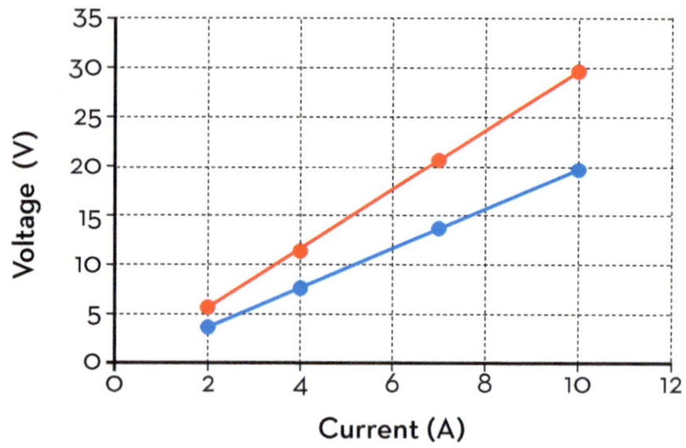

Look at the graph. It represents the current that circulates through a circuit according to the voltage. The blue line represents one circuit and the red line represents a different circuit.

- Which of these circuits needs more voltage to have a current of 7 A?

From the description of current intensity, voltage and resistance given in section 2, which circuit has a higher electrical resistance?

WORKSHEET

Download this unit's worksheet and do activities 3 and 4.

ELECTRICITY 2.0

You can practice Ohm's law in the following link:

goo.gl/piyMia

Ohm's law is the mathematic expression that quantifies the relationship between the three electrical magnitudes described:

The intensity of the current that circulates through a circuit, the resistance of this circuit and the voltage are related through **Ohm's law:**

$$V = I \cdot R$$

SOLVED EXAMPLE

1. In the electrical device in the figure, the electrons are 'driven' by a voltage of 3 V. The device provides a resistance to the passage of current of 6 Ω. What is the intensity of the current? What current will there be if the resistance is doubled? What if the voltage is doubled?

Solution

Using Ohm's law, isolate the current intensity and then enter the numerical values:

$$V = I \cdot R \qquad I = \frac{V}{R} = \frac{3\ \text{V}}{6\ \Omega} = 0.5\ \text{A}$$

If we double the resistance, the electrons will have greater opposition to their circulation. Therefore, the intensity should decrease. On the contrary, if we increase the voltage, that is, the impulse of the electrons, the intensity increases. So:

$$I = \frac{V}{R} = \frac{3\ \text{V}}{12\ \Omega} = 0.25\ \text{A} \qquad I = \frac{V}{R} = \frac{6\ \text{V}}{6\ \Omega} = 1\ \text{A}$$

2.2. Electrical conductors and insulators

Not all materials allow electric current to pass through in the same way. Their different behaviour allows us to use them in electrical devices to control the flow of electrons.

THINK AND REASON

Will the light bulb light up when we connect the cables to the wooden pencil, the plastic pen, the nail or the key?

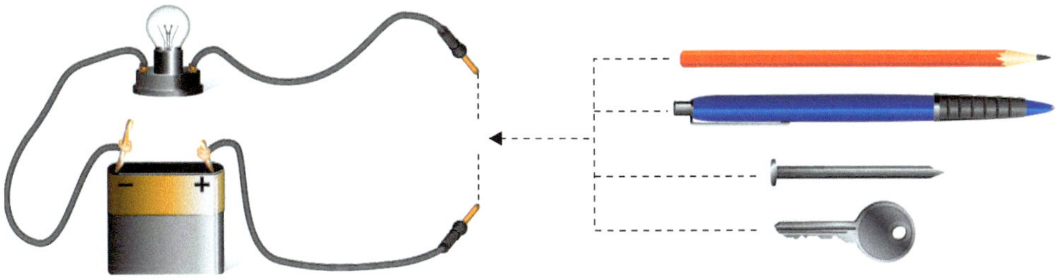

We produce an electric current when we apply a voltage between two points of a material that is an **electrical conductor**. This means that its atomic structure allows the current of electrons to pass through it.

There are many materials whose atomic structure blocks the passage of the electron current. These materials are called **electrical insulators**. Both materials are incredibly important in our daily lives.

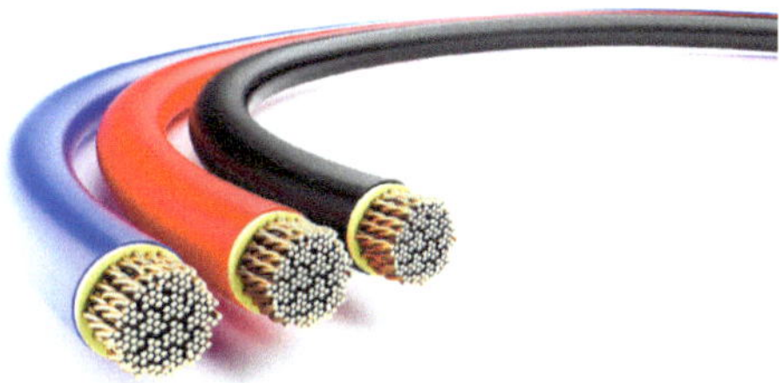

Electrical cables are made of copper and aluminium. These are excellent electrical conductors and they can be shaped easily. The cables are covered in plastic, an electrical insulator. This prevents coming into contact with the current when touching the cable.

The handles of tools, electrical tape and socket plates are made of plastic. This is the most commonly used insulating material.

ELECTRICITY 2.0

Open the circuit simulator that we looked at in Electricity 2.0 of section 1.

You can go directly to:

goo.gl/sZ3HXq

Build a circuit with a light bulb, a battery and conductors. Use the bag of objects to put other materials in the circuit and learn the difference between conducting and insulating materials.

WORKSHEET

Download this unit's worksheet and do activities 5 and 6.

ACTIVITIES

4. The electrical resistance of a copper cable 1 km long and 1 cm in diameter is 0.055 Ω. If the same cable were made of steel, its resistance would be 2.30 Ω. What material is a better electrical conductor, copper or steel? Compare your reasoning with your partner.

5. You can see glass rings holding the cables on lamp posts. Search on the internet for their function and the reason why they are made of glass. Write a brief report about this and then explain it to your classmates in an oral presentation. You can use some pictures to support your presentation.

3 Electric circuits

An **electric circuit** is a set of electrical elements that are connected together in a closed form. It allows us to generate, transport and use the electrical energy through its transformation into another type of energy.

WORKSHEET

Download this unit's worksheet and do activities 7, 8 and 9.

THINK AND REASON

Look at the figure and answer:

- What would happen if the pump did not exist?
- What does the turbine turn the potential energy of the water into?

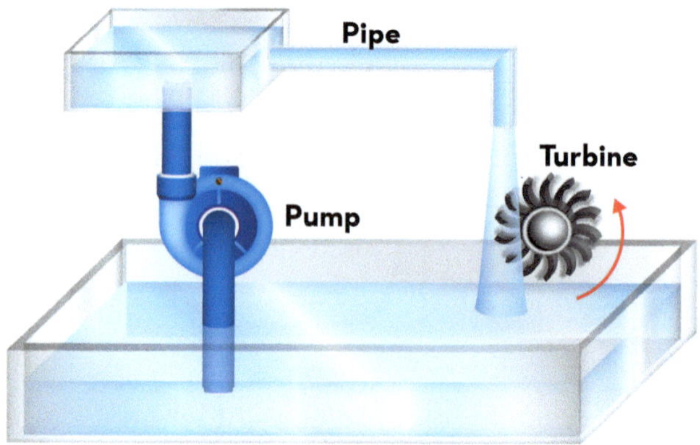

3.1. Elements of an electric circuit

All electric circuits should be made up of one or more generators, conducting cables, receiving elements and control elements.

Receiving elements like light bulbs transform electrical energy into another type of energy, in this case, light energy. All the electrons have the same energy when passing through the receiver. Therefore, the greater the current, the more brightly the light bulb will shine.

Conducting cables allow the current to pass, but oppose with a resistance that depends on the material they are made of, its section and its length. In a simple circuit, this resistance is insignificant.

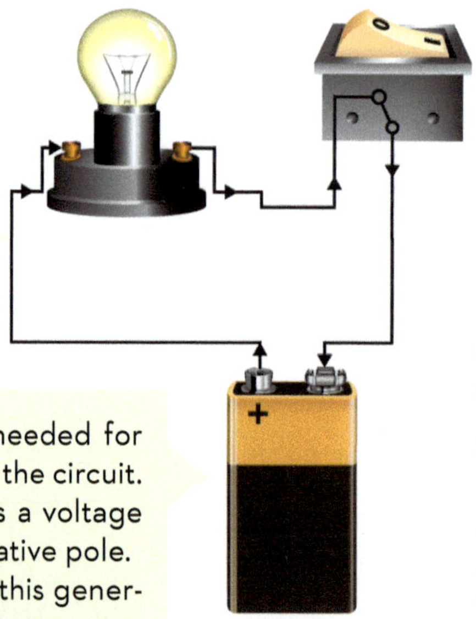

The switch is the **control element.** If it is open, it blocks the passage of current. The control elements are indispensable for the energy of the generator to be consumed only when necessary.

The **generator** produces the voltage needed for the electric current to circulate through the circuit. For example, a 4.5 V generator creates a voltage of 4.5 V from its positive pole to its negative pole. Another way of interpreting this is that this generator is able to deliver 4.5 joules of energy to one coulomb of charge before consuming itself.

There are also **protection elements** in circuits, like fuses. These protect the elements of the circuit from possible overloads. Other elements in circuits are **electronic elements.** We will study these at the end of the unit.

3.2. Functioning of an electric circuit: connection in series

There are two main ways of connecting the elements of a circuit together. The first is to connect them **in series**. This means that the current must go through all the elements, one after another.

CLASSROOM LABORATORY

2. You need identical light bulbs with their lamp holders, 4.5 V batteries, switches and conducting cable.

   The brightness of a light bulb depends on the current that goes through it and its electrical resistance. We will use identical light bulbs so that the brightness will indicate the current.

   First, build circuits 1 and 2. Connect the second one in series. When you finish building, **do not disassemble the circuit**; you will need it for the next page.

Circuit 1

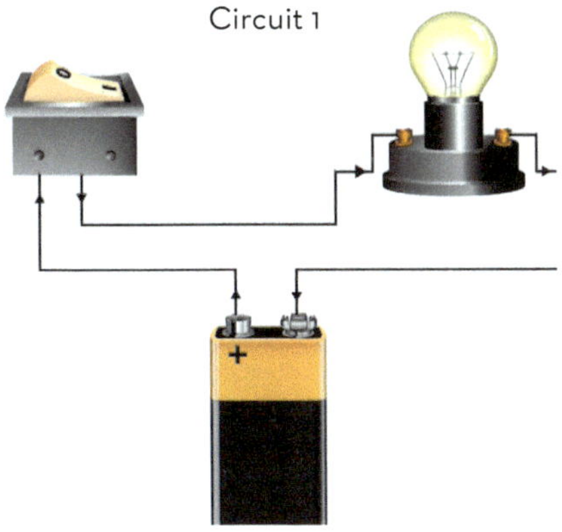

Circuit 2

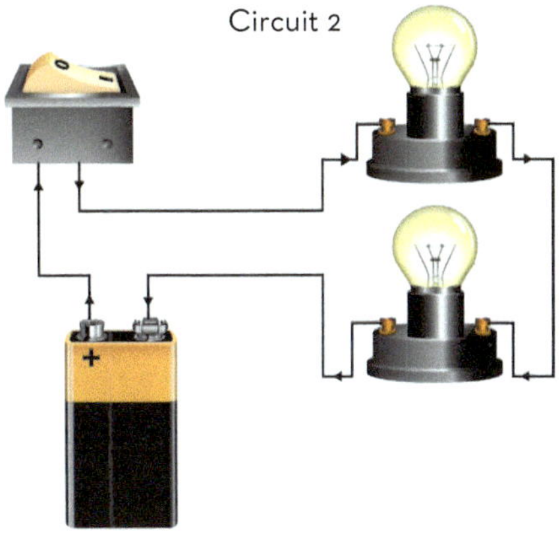

- Do the two light bulbs in circuit 2 have the same brightness? Therefore, does more current go through one of the two light bulbs of this circuit?
- Compare the brightness of the light bulb in circuit 1 with the light bulbs in circuit 2. Is there more current going through the light bulbs in circuit 1 or circuit 2?
- As the batteries of both circuits are equal (both 'push' the electrons the same), use your answer from the previous question and answer the following: Is the total resistance (opposition to the passage of current) of two light bulbs in series greater than or less than that of only one light bulb?

To finish, build circuit 3, connecting two batteries in series. Connect the positive pole of one battery with the negative pole of another.

- Compare the brightness of the light bulbs in circuits 2 and 3. Does the same current circulate through these light bulbs? The resistance of circuits 2 and 3 is the same. Therefore, is the total voltage of two generators in series greater than, equal to or less than that of only one generator?

IMPORTANT !

We represent the most important elements of a circuit with symbols. These symbols facilitate drawing circuits. The most important symbols are:

Symbol	Element
	Battery
	Light bulb
	Switch
—	Conductor
	Button
	Motor

Circuit 3

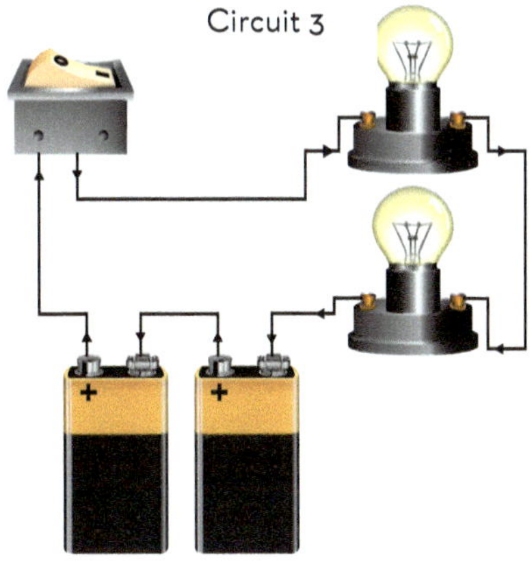

3.3. Functioning of an electric circuit: connection in parallel

The second way of connecting the elements of a circuit together is **in parallel.** We make this connection by connecting the elements in such a way that the electric current is divided between them.

WORKSHEET

Download this unit's worksheet and do activities 10 and 11.

CLASSROOM LABORATORY

3. Now you are going to build two circuits to observe how receivers and generators connected in parallel behave. For the new circuits, besides the batteries, cables, identical light bulbs and lamp holders, you will need terminal strips to make the divisions of the cables.

Build circuit 4 by connecting two light bulbs in parallel and one in series with them.

- Compare the brightness of the two light bulbs in parallel in circuit 4. Is one light bulb brighter? Therefore, does the same current go through them?
- Compare the brightness of the light bulbs in parallel with the light bulb in series with them. Can you explain experimentally that the current is divided between the two branches in parallel?

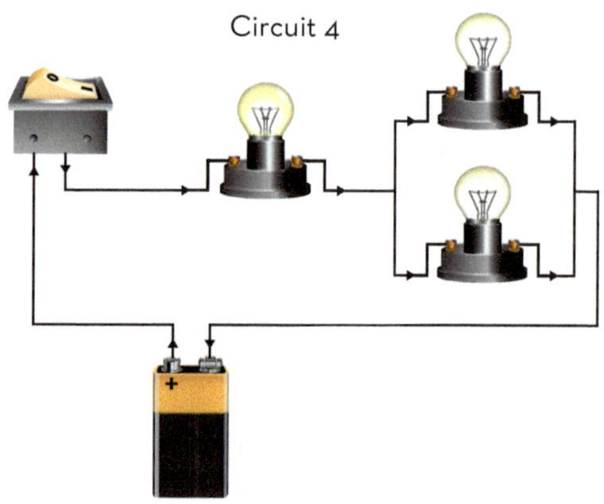

- Unscrew one of the light bulbs in parallel in circuit 4 and then unscrew the light bulb in series. What happens to the rest of the light bulbs in each situation? Explain your answer using the definition of electric current.

Now build circuit 5. You must connect two generators in parallel, and then connect the light bulb to one branch of the circuit and the switch to the other branch.

- Compare the brightness of the light bulbs in circuits 1 and 5. Does the same current circulate through the two light bulbs or is it greater for one of them?
- The electrical resistance of circuits 1 and 5 is the same. Using the answer from the previous question, answer the following: Is the voltage created by two generators in parallel greater than, less than or equal to that created by only one generator?

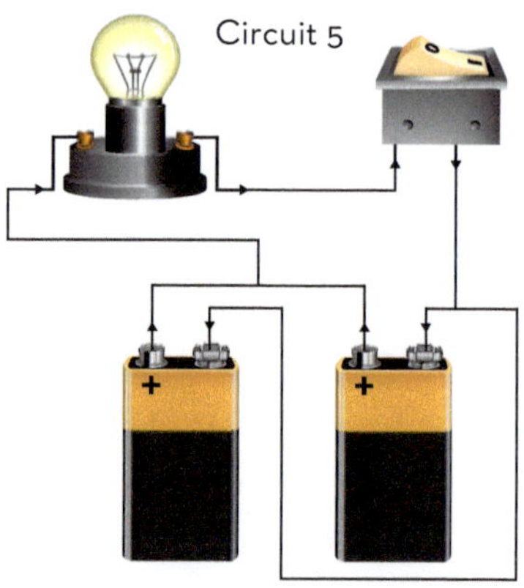

3.4. Solving circuits by using Ohm's law and virtual applications

On the previous pages, in an experimental way, you have reached various qualitative conclusions on the functioning of electric circuits. On the following pages we want to give a mathematical form to these qualitative conclusions. To do this, we will use Ohm's law and a virtual circuit simulator that you will find at **phet.colorado.edu/en/simulation/legacy/circuit-construction-kit-dc.** Select *Physics*, then *Circuit Construction Kit (DC Only)*. Direct link: **goo.gl/RjQa9E**

A. Association of resistors in series

With the following example, we will see that:

The same current circulates through two elements connected in series. The equivalent resistance of two resistors in series is the sum of these resistors:

$$R_{eq} = R_1 + R_2$$

SOLVED EXAMPLE

2. Use the circuit simulator that we indicated and build the circuit in the figure. One of the light bulbs has 10 Ω of resistance and the other has 20 Ω of resistance. The battery provides a voltage of 9 V.

Does the same current circulate through the two light bulbs? What is the equivalent resistance of the two light bulbs in series?

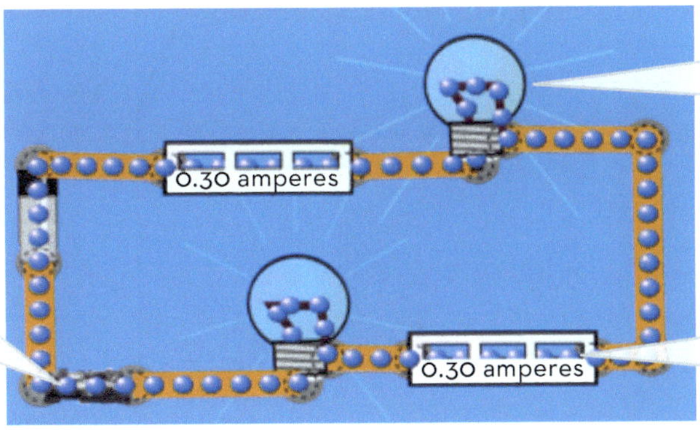

To calculate the equivalent resistance of the two light bulbs in series, we will apply Ohm's law. As the battery provides 9 V and the ammeters show that the current is 0.30 A, the resistance of the set of two light bulbs is:

$$R = \frac{V}{I} = \frac{9\ \text{V}}{0.30\ \text{A}} = 30\ \Omega$$

You can see that this value corresponds to the sum of the individual resistors like we wanted $R_{eq} = 10\ \Omega + 20\ \Omega = 30\ \Omega$.

B. Association of resistors in parallel

By using the virtual circuit simulator, we want to see that:

> The electric current is divided between the two branches of a circuit connected in parallel.
>
> The equivalent resistance of two resistors in parallel is calculated with:
>
> $$\frac{1}{R_{eq}} = \frac{1}{R_1} + \frac{1}{R_2}$$

SOLVED EXAMPLE

3. Use the circuit simulator and build the circuit in the figure. The resistance of the light bulb in the upper branch is 10 Ω and the one in the lower branch is 20 Ω. The battery provides a voltage of 9 V.

What is the voltage of the two branches in the circuit? How is the current divided between the branches? What is the equivalent resistance of the two light bulbs connected in parallel?

From the values shown by the ammeters, we can see that the incoming electric current is divided between the two branches of the circuit:

1.35 A = 0.90 A + 0.45 A

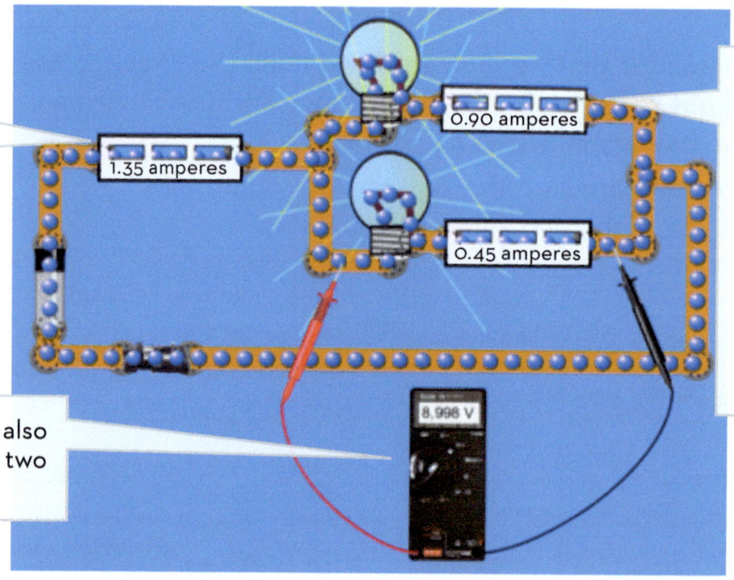

The voltmeter of the simulator also tells us the voltage between any two points of the circuit.

By applying Ohm's law, calculate the voltage in both branches:

- Upper branch:

  $V = 0.90\ \text{A} \cdot 10\ \Omega = 9\ \text{V}$

- Lower branch:

  $V = 0.45\ \text{A} \cdot 20\ \Omega = 9\ \text{V}$

We see that both branches in a parallel circuit have the same voltage.

Use Ohm's law to calculate the equivalent resistance of the two light bulbs. The electric current outside the parallel branches is 1.35 A and the voltage created by the battery is 9 V. Therefore:

$$R = \frac{V}{I} = \frac{9\ \text{V}}{1.35\ \text{A}} = 6.67\ \Omega$$

This value observes the relationship indicated for parallel resistors:

$$\frac{1}{R_{eq}} = \frac{1}{10} + \frac{1}{20} = 0.15 \rightarrow R_{eq} = \frac{1}{0.15} = 6.67\ \Omega$$

C. Association of generators

Finally, we will see that:

- Two generators connected in series provide a voltage equal to the sum of the voltages of both generators.
- Two generators connected in parallel provide the same voltage as only one of the generators.

WORKSHEET

Download this unit's worksheet and do activities 12 and 13.

SOLVED EXAMPLE

4. Build the following two circuits with 9 V batteries with a 10 Ω resistance and observe the values provided by the simulator. What is the voltage of two generators in series and in parallel?

This light bulb is brighter than that of the other circuit because the current going through it is greater.

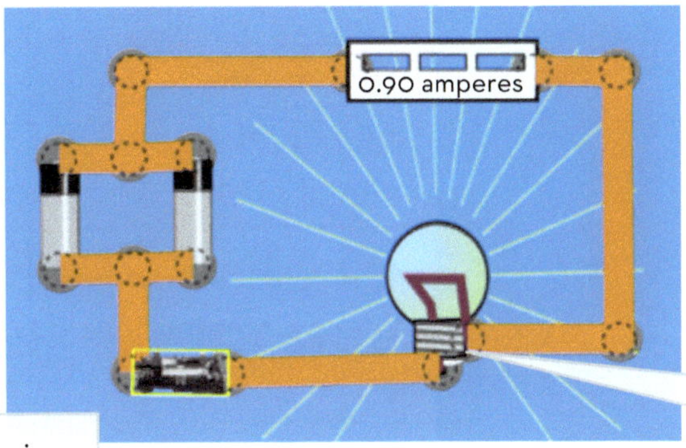

By applying Ohm's law, we obtain the voltage of the generators in parallel: $V = 0.90\ A \cdot 10\ \Omega = 9\ V$, the same as only one of the generators.

By applying Ohm's law, we obtain the voltage of the generators in series: $V = 1.80\ A \cdot 10\ \Omega = 18\ V$, corresponding to the sum of the voltages of the two batteries.

ACTIVITIES

6. A circuit is powered by two 9 V generators in series and has a resistance of 12 Ω. How much electric current circulates through the circuit? Compare your results with your partner.

7. A circuit is powered by two 9 V generators in parallel and has a resistance of 12 Ω. How much electric current goes through the circuit? Explain it to your partner.

8. Will the generators of activity 6 or activity 7 be used up first?

9. Calculate the electric current that circulates through a circuit powered by a 9 V generator with three light bulbs in series of 2 Ω each. Share your results with your partner.

10. Calculate the electric current that circulates through each branch of a circuit made up of one 9 V generator and two light bulbs in parallel, with resistors of 2 Ω and 4 Ω.

11. The generator of a circuit is substituted with another generator that provides double the voltage between its terminals. What happens to the current? Discuss your conclusions with the rest of the class.

4 Electrical machines

Electric circuits are the basis of all the electrical machines around us. They transform electrical energy into heat, movement, light or sound. Until now, we have been working with circuits where electric energy is transformed into light. Now we are going to see how electric energy can be transformed into movement.

EXPERIMENT A very simple electric motor

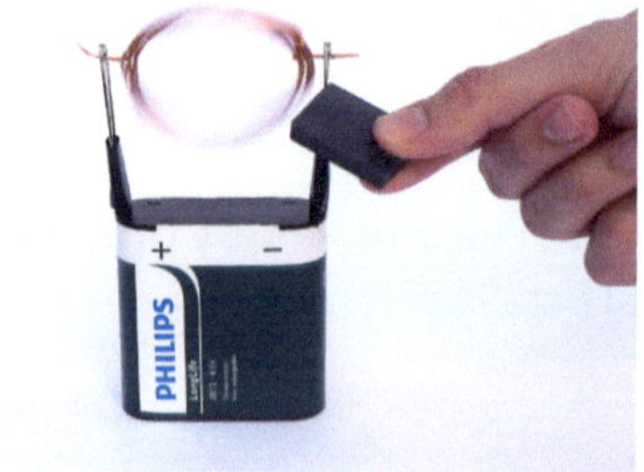

1. You are going to build a simple electric motor. You need a battery (4.5 V), enamelled copper wire, a magnet, two safety pins, electrical tape and sandpaper.

   Make a small coil with the copper wire, leaving a few centimetres uncoiled at each end to form the rotation axis. These ends must be aligned and straight and you must remove the enamel covering them with the sandpaper. Join the two safety pins vertically to the two terminals of the battery with electrical tape. Finally, put the straight ends of the coil in the holes of the safety pins, bring the magnet close and the coil will begin to turn on its axis.

   - Identify the generator, the conductors and the receiving element in the circuit of your electric motor.

WORKSHEET

Download this unit's worksheet and do activities 14 and 15.

Electric motors are elements that transform electrical energy into rotational kinetic energy. Heaters are devices that transform electrical energy into thermal energy.

Electric motor: present in mixers, washing machines, blenders, vacuum cleaners, etc.

When electric current circulates through a conducting coil located between magnets, the coil begins to turn. This turning is transmitted to a shaft that is joined to the arm of a mixer.

Electric heater: present in water heaters, ovens, toasters, glass ceramic hobs, etc.

The heaters in water heaters and glass ceramic hobs have elements with a lot of electrical resistance and high thermal conductivity. This high resistance makes the electrical energy transform into thermal energy.

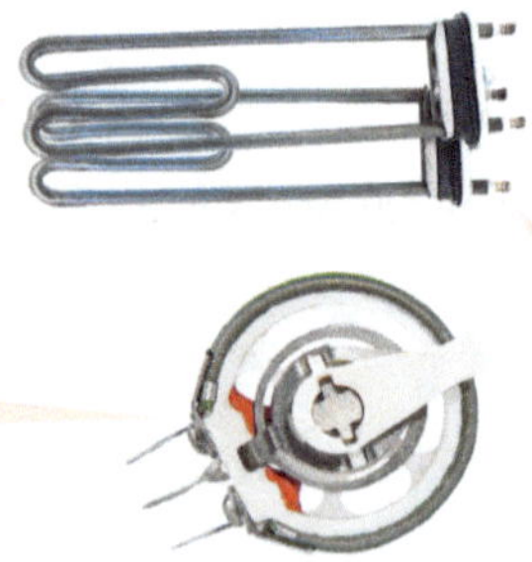

5 How is the electrical energy in our homes produced?

Currently, human beings are able to transform rotational kinetic energy, the electromagnetic energy of light and chemical energy into electrical energy. It is not possible to transform thermal energy or sound energy directly into electrical energy.

- The **photoelectric effect** is a physical phenomenon. When light with the right characteristics shines on metals or other materials, they emit electrons. These can be canalised into an electric current. This is used in photovoltaic solar panels.
- **Electrochemical reactions** transform chemical energy into electrical energy. They are used in the batteries of cars, mobile phones and pacemakers.
- The phenomenon of **electromagnetic induction** involves the rotation of a conducting coil in the presence of a series of magnets. This creates a voltage in the coil, which creates an electric current. Electromagnetic induction is based on the opposite effect of the electric motor.

WORKSHEET

Download this unit's worksheet and do activities 16 and 17.

Electrical power plants are industrial facilities where rotational kinetic energy from different sources is transformed into electrical energy. The key element of these plants is an **electric generator**, like that shown in the following figure.

ACTIVITIES

12. Look for information on how a bicycle light works with an electrical dynamo. Explain the origin of the electrical energy produced and the way it is transformed into light. Prepare a brief presentation and show it to your class.

13. Find information on the current efficiency of transforming electromagnetic energy into electrical energy from a solar panel. Prepare an elevator pitch (short speech in the time that it takes to ride up an elevator) about the future possibilities of solar panels.

5.1. Generation, transport and storage of electrical energy

Transporting electrical energy from the electric power plant to the end user is a complex process. This is because they must reduce energy losses to a minimum and they must deliver the current with the right voltage for each type of user.

Nuclear power plants use the enormous amount of thermal energy released in a nuclear fission reaction to produce water vapour. This vapour moves the blades of a turbine. The turbine is connected to a generator that transforms its rotational movement into electrical energy.

Thermal power plants burn natural gas or coal. The energy from this combustion is used to produce water vapour. Like in the nuclear power plant, this causes the rotation of a turbine to produce electrical energy in the generator.

Wind power plants have a large number of wind turbines. Each one of them transforms the kinetic energy of wind into energy to rotate the blades of the turbine. Inside, an electric generator transforms this rotational energy into electrical energy.

Hydroelectric power plants use water that falls from an elevated dam to the turbine and the generator, located at a lower level. The stored potential energy of the water is transformed into rotational kinetic energy and then finally into electrical energy.

Part of the electrical energy produced is used to pump the water back to the reservoir. This makes it one of the few present-day systems of **electrical energy storage.**

ACTIVITIES

14. Together with your partner, do research on how many major thermal, hydroelectric and nuclear power plants there are in Spain. Where are they located? Are there more in the north or in the south of the country? Explain the reasons for this distribution.

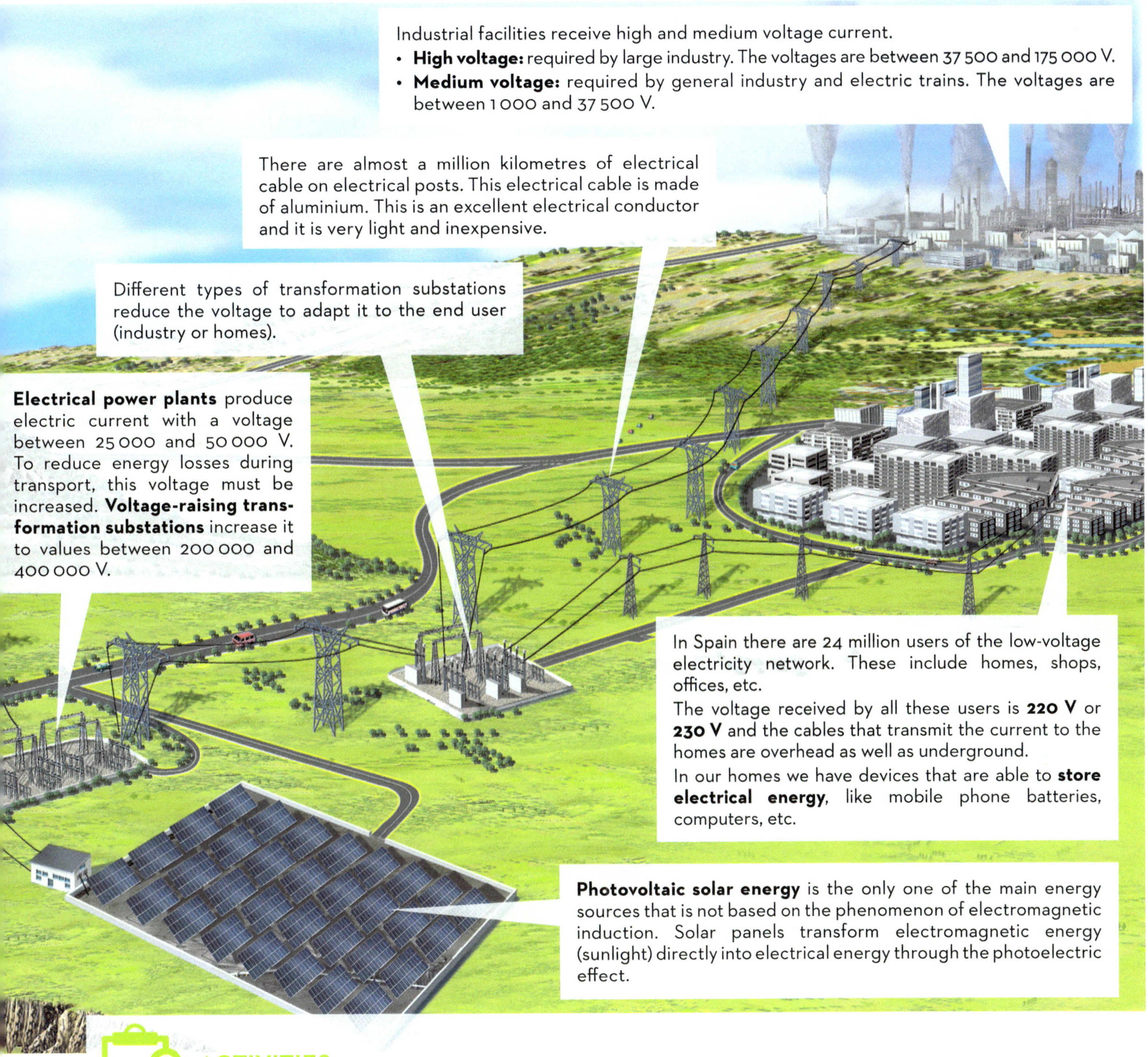

Industrial facilities receive high and medium voltage current.

- **High voltage:** required by large industry. The voltages are between 37 500 and 175 000 V.
- **Medium voltage:** required by general industry and electric trains. The voltages are between 1 000 and 37 500 V.

There are almost a million kilometres of electrical cable on electrical posts. This electrical cable is made of aluminium. This is an excellent electrical conductor and it is very light and inexpensive.

Different types of transformation substations reduce the voltage to adapt it to the end user (industry or homes).

Electrical power plants produce electric current with a voltage between 25 000 and 50 000 V. To reduce energy losses during transport, this voltage must be increased. **Voltage-raising transformation substations** increase it to values between 200 000 and 400 000 V.

In Spain there are 24 million users of the low-voltage electricity network. These include homes, shops, offices, etc.

The voltage received by all these users is **220 V** or **230 V** and the cables that transmit the current to the homes are overhead as well as underground.

In our homes we have devices that are able to **store electrical energy**, like mobile phone batteries, computers, etc.

Photovoltaic solar energy is the only one of the main energy sources that is not based on the phenomenon of electromagnetic induction. Solar panels transform electromagnetic energy (sunlight) directly into electrical energy through the photoelectric effect.

ACTIVITIES

15. The picture shows the electrical distribution lines in Spain that transport electrical energy between 200 and 400 kV. The orange circles show the areas of high concentration of transformation stations, for increasing as well as decreasing voltage. Based on the answer from activity 14, explain why there are more lines in the north than in the south, what the empty space in Castile-La Mancha means and why there is an orange circle on the main capital cities. Prepare a short video with the result of the activity and play it for the class.

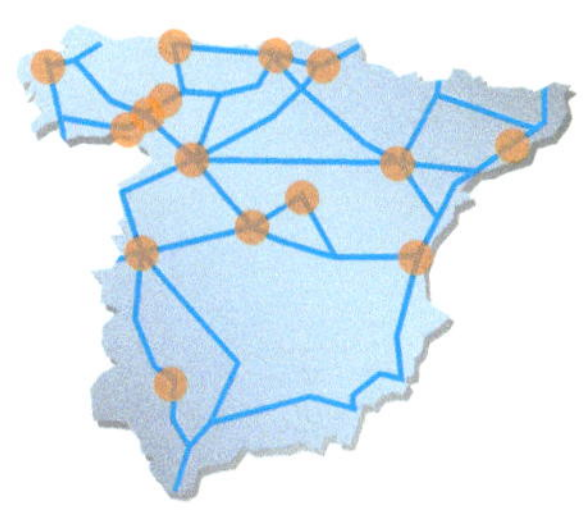

5.2. Electrical installation in our homes

WORKSHEET

Download this unit's worksheet and do activities 18, 19 and 20.

We can understand the electrical installation in our homes by comparing it to a large electric circuit.

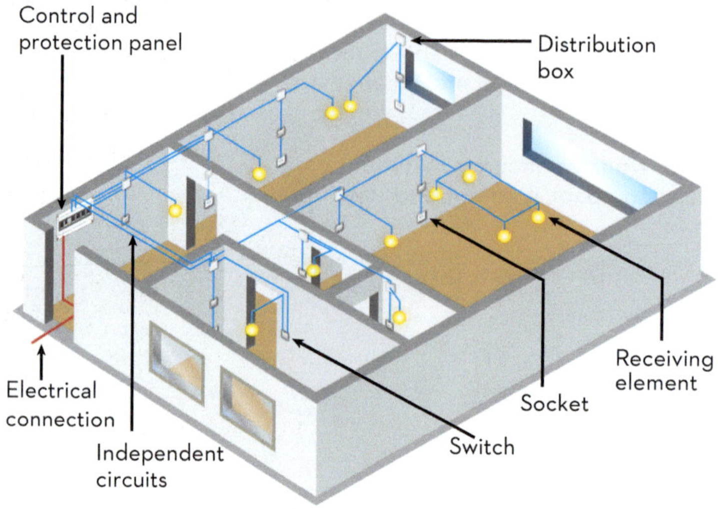

Generator

Our homes do not have their own generators. The current arrives with the correct voltage through the exterior lines. This current enters our homes through a point known as the **electrical connection**.

All the sockets of a home are points where we can connect any device to supply it with 220 V electric current.

Protection elements

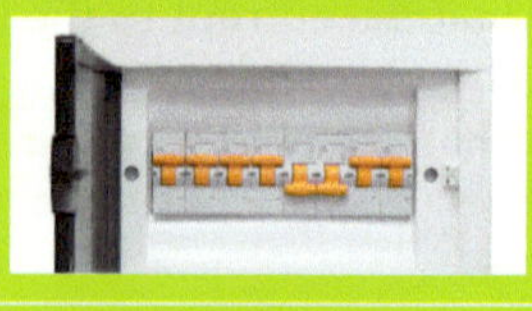

The electric current enters our homes through the **control and protection panel**. This panel has several switches. These protect our homes from electrical overloads.

Control elements

Our homes have **distribution boxes**. These are where the current branches to reach all the points of our homes. There are also **switches** to control the passage of current and turn lights on or off.

Conductors

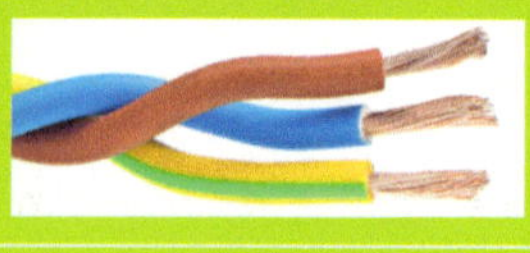

There are three types of cables in our homes. The **phase** cable is brown, black or grey and is where the current enters. The **neutral** cable is blue and is where the current leaves. The **earth** cable is green and yellow and its function is to evacuate current leaks to a safe place.

Receiving elements

These are the elements where the electrical energy is transformed into another type of energy: light bulbs, electrical devices, electric heating, etc. The energy consumed in a home is quantified in the electricity bill.

In recent years, the increase in the cost of electric current together with an increase in environmental protection and technical possibilities have allowed more and more people and companies to consider producing their own electric current. This is called **electrical self-consumption.**

5.3. Labelling on electrical devices

The **labels** on all electrical devices include a series of **symbols** and **abbreviations** with important information on their characteristics. The most important ones are:

Manufacturer, type of device and **model.**

Safety warnings to prevent accidents when using the device.

International certifications: The most important is the CE certification. This is indispensable for selling the device in the European Union. It informs the consumer that the device meets the standards of quality and safety for the European Union.

The **wave symbol** means that the device uses alternating current. The **power voltage** should be between 220 and 240 V. This is valid for Spain, but not for the United States, for example, where they use 120 V. 179 W is the power consumed by the device.

This indicates that the device has a **double insulation protection** on all its metal parts. This is important in devices that are used outside or possibly come into contact with water, like in the bathroom.

This indicates that the device should not be thrown in the rubbish. It should be taken to a collection point because it was designed so that most of its materials can be recycled.

EXPERIMENT

2. Look in your home for labels on a television set, a refrigerator and a mobile phone (many times they are located under the battery).

   In direct current, the electrons always circulate in the same direction. In alternating current, the electrons change the direction of their movement tens of times per second. The symbols for these currents are shown in the figure.

Alternating current	Direct current
~	⎓

   Which of the three devices you have looked at use alternating current and which use direct current? Which of them requires more power? Do they all have the CE symbol? Should you take them all to a collection point? Share your results with the class.

ACTIVITIES

16. The simplest household self-consumption installation has one or more solar panels connected to the electrical installation of a home. Is this system useful for producing the electrical energy required by the washing machine in the home? Is it useful for producing the electrical energy required by the refrigerator? What would be needed to complete the installation described and supply the home with all the energy needed?

17. Do research with your partner about *Tesla Powerwall* batteries and prepare an oral presentation to convince your classmates to install a self-consumption system in their homes linked to one or more of these batteries. Use economic and environmental arguments.

6 The electronic revolution

WORKSHEET

Download this unit's worksheet and do activities 21 and 22.

Electronics is the discipline in charge of designing devices that use electric currents for generating, transmitting or storing information.

THINK AND REASON

The figure shows a simple electronic circuit. It functions as an alarm when the cable is broken. The element named BC 108 is a transistor.

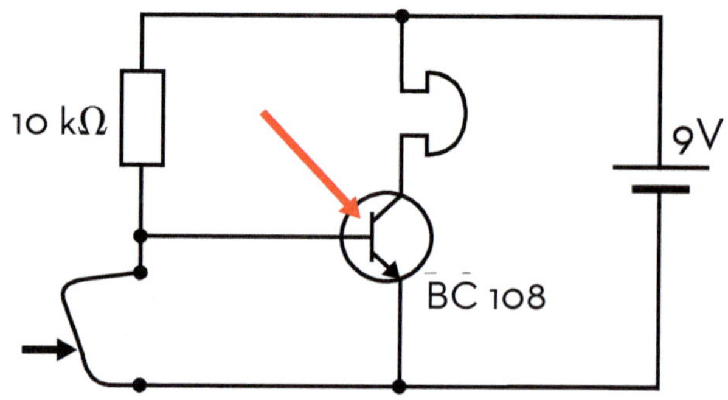

A transistor located where it is in the figure acts as a switch. It stays open when there is a large electrical resistance. It closes and lets current pass through when the current that reaches the point of the transistor marked with a red arrow is great enough. The symbol that comes after the transistor is a bell. It sounds when electric current passes through it.

Why does the bell only sound when the cable marked with a black arrow is broken? Where do you think we can find circuits like this one?

- The main electronic devices are **resistors**, **variable resistors**, **diodes** and **transistors**. Their functions are shown in the following examples.
- The following circuit is a **heat detector**. The electronic components transform the information (low temperature) into an electric current that sets off an alarm. In this case the LED light turns on.

LED diode: A diode is a device that only allows current to pass through in one direction. An LED diode emits light when current passes through it.

NTC thermistor: The resistance value increases when the temperature decreases.

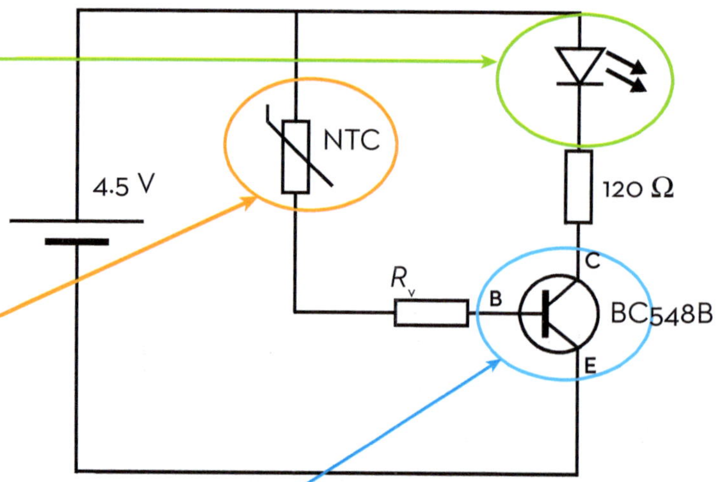

Transistor: Depending on the current it receives for each one of the three connections, the transistor acts as an open or closed circuit, permitting or blocking the passage of current between two of its connections.

The current leaves through the generator. If the temperature is high, the NTC resistance is low and the current that reaches the transistor is enough to allow its passage. This causes the light to turn on, informing of the high temperature.

If the temperature is low, the NTC resistance is high and the current that reaches the transistor is insufficient. In this case, the transistor acts as an open circuit and the current does not circulate through the LED.

- The following example is a circuit that reacts when the light intensity that reaches the system is less than a given value. Therefore, it is a darkness detector circuit.

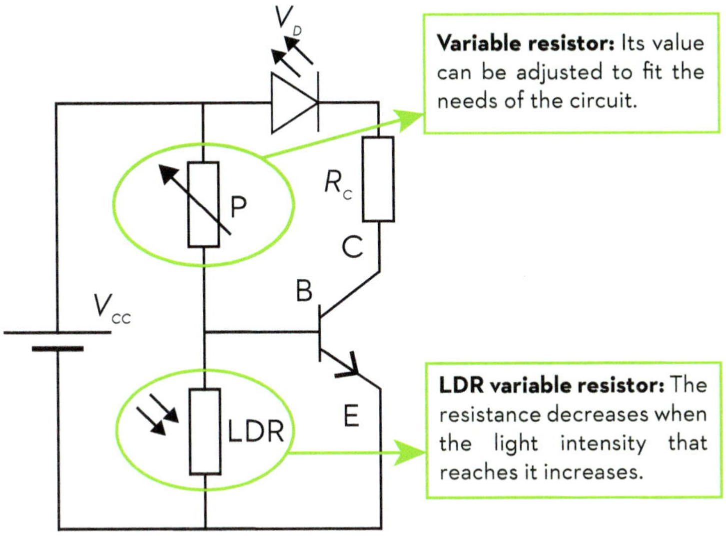

The current leaves through the generator. If the light intensity is high, the LDR resistance is low and the current goes through this resistor instead of going to the transistor. In these conditions, the transistor blocks the passage of electric current to the LED diode.

When the light intensity is low, the LDR resistance is high. Therefore, the current mainly goes through the transistor. In these conditions, the transistor permits the passage of current to the LED diode. The light turns on to inform of the low light intensity.

6.1. Integrated circuits

The previous circuits show the great ability of electronics to obtain and transmit information through electric currents. Since the mid-twentieth century, engineers have an electronic tool that is so powerful it has radically transformed our lives: the **integrated circuit.**

An **integrated circuit, chip or microchip** is a structure that has a few square centimetres or millimetres of area. Electronic circuits are printed on them and there may be millions of transistors on this surface.

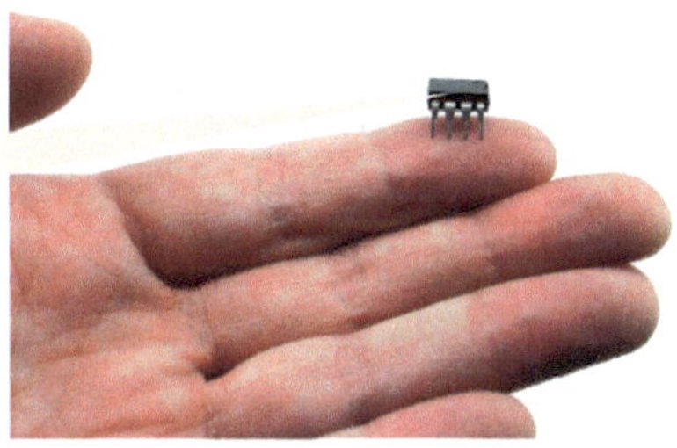

With millions of transistors it is possible to manage very complex information, like that on our mobile phones and computers. All our information society depends on integrated circuits. They are indispensable thanks to their small size, incredible efficiency and low cost.

ACTIVITIES

18. The following electronic circuit is a cold detector. Compare it with the heat detector explained on the previous page and describe how it works with the help of your partner.

19. The first computers used vacuum valves instead of transistors. A vacuum valve has a height between 10 and 15 cm. What characteristics of our modern lives would not be possible if human beings had not discovered the transistor? Discuss this with your partner.

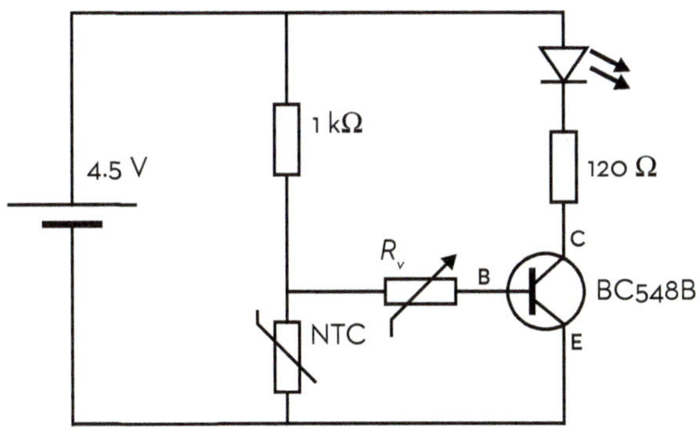

MIND MAP

Download and print the worksheet you will find in the OLC to complete the mind map of the unit.

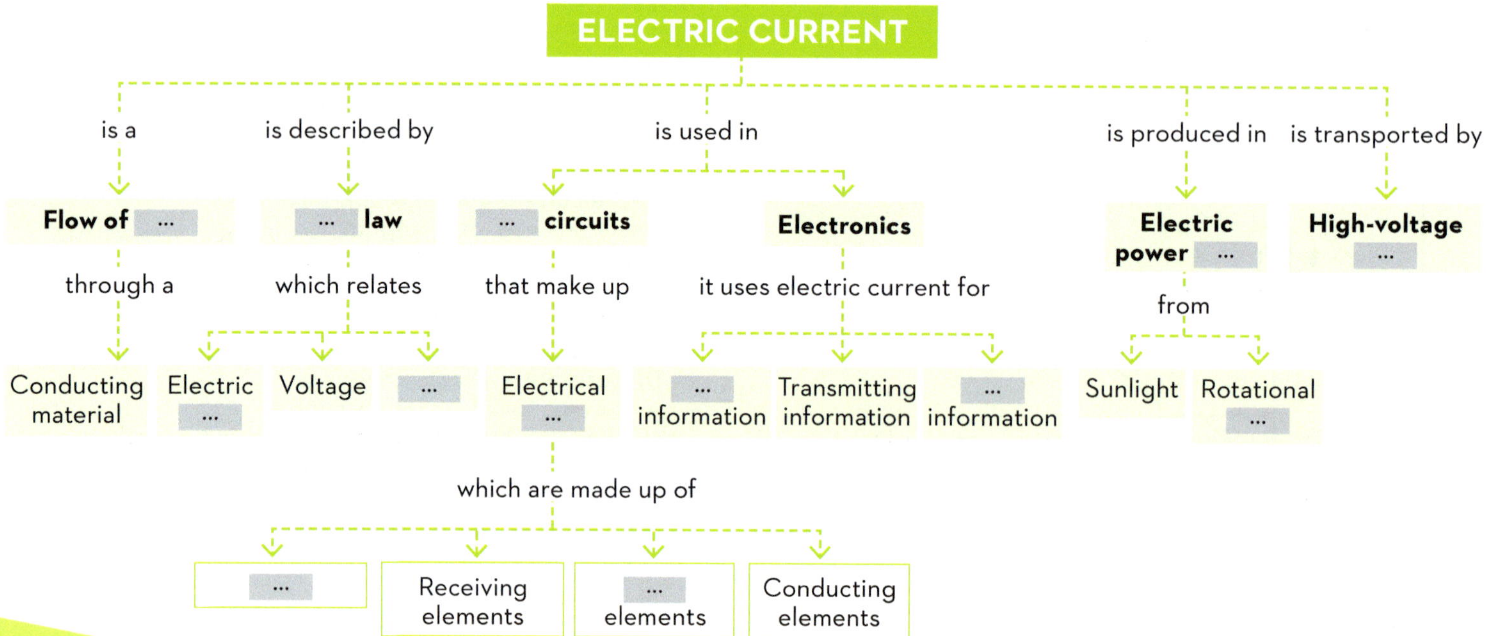

LOOK AROUND YOU

The war of the currents

He arrived in New York in 1884. He had a letter of recommendation: 'I know two great men, and you are one of them. The other is the young bearer of this letter'. The recipient of this letter was the already famous Thomas Alva Edison, the father of the light bulb. The other 'great man' was Nikola Tesla, a young Serbian who came from Edison's Parisian-based office.

When he arrived in New York, Tesla asked where Edison's offices were. There he went to talk with his future boss and he left the office with a job. But there was little harmony between the two men. Soon their incompatibility ended in open conflict. Edison defended an electrical industry based on direct current. His electrical distribution system, the first in the world, started in 1882 with 110 V of direct current and 59 clients in Manhattan. Tesla, on the other hand, believed in a model based on alternating current. The conflict between them went down in history as 'the war of the currents'. Tesla won with his model, which was much more efficient. Thanks to him, pressing a switch lights up our homes today. The world owes a lot to the enigmatic Nikola Tesla.

ABC newspaper. 23/04/2012

Questions

a) Look up information and explain to the class why alternating current is much more used than direct current now. Which of these currents do we have in our homes?

b) Edison did not hesitate to gather journalists to show them the dangers of alternating current, by applying shocks to dogs and cats he had collected from the streets. Do you think the progress of science and technology is affected by personal animosities and methods like those used by Edison?

c) Write five true/false statements about the text and give them to your partner to check his/her understanding. Correct them together.

LABORATORY WORK

Use of the multimeter

Objective

The objective of this practice is to measure the most significant values of voltage and current in an electric circuit.

Material

A 4.5 V battery, three light bulbs, a switch, conducting cables and a digital multimeter.

Procedure: work in pairs

The multimeter is a device that measures the current and voltage in several places in a circuit.

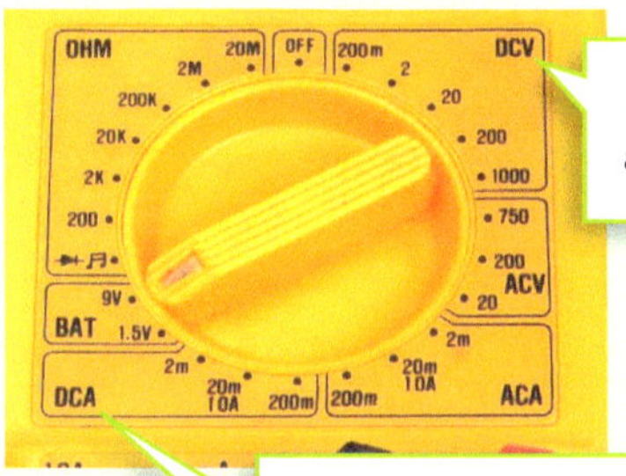

Put the dial in the DCV area to measure voltage.

Put the dial in the DCA area to measure current.

Start with the largest values on the scale and go down the scale, little by little, adjusting the measurement.

To measure the voltage between two points:

Touch the leads of the multimeter to the two points where you want to know the voltage. On a generator, touch the two terminals; on a light bulb, touch the two connections.

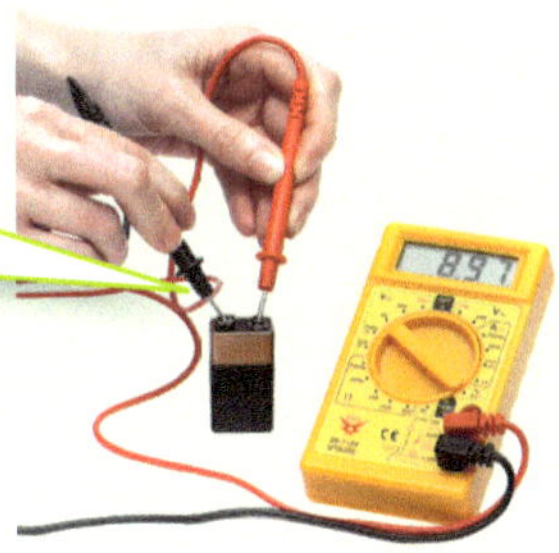

To measure the current, connect the multimeter in series at the point of the circuit where you want to measure the current.

You must build the following circuit with the help of your partner, showing the location of the multimeter for measuring the voltage in light bulb 1 and its location for measuring current in light bulb 2:

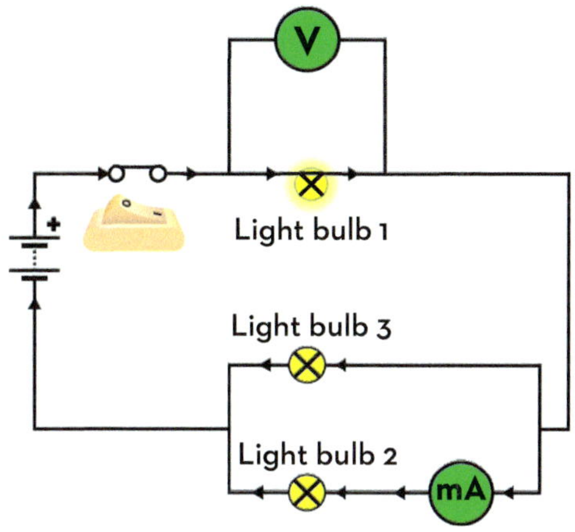

You must measure the current that goes through each of the three light bulbs and the voltage between the entrance and exit of each of the three light bulbs.

Questions

a) Make a table showing all the values measured with the multimeter.

b) See that the current going through light bulb 1 is divided between light bulbs 2 and 3. Explain this fact to your partner according to the definition of electric current.

c) See that the total voltage that the generator supplies (4.5 V) is equal to the sum of the voltages of light bulb 1 and the set of light bulbs 2 and 3.

FINAL ACTIVITIES

Basic activities

1. A student makes the following statement: 'The difference between a material that is a good electrical conductor and one that is a bad conductor is equivalent to the difference between having a race on asphalt or on desert sand'. Do you agree? Explain your answer.

2. Make a list with six electrical devices and indicate what type of energy each one of them transforms electrical energy into. Share your list with the rest of the class and make a poster with all the devices you have listed. Classify them by the types of energy they generate.

3. Complete the following table in your notebook:

Current (A)	Voltage (V)	Resistance (Ω)
9	4.5	...
...	9	3
10	...	0.25

4. Calculate the total resistance of the following circuits:

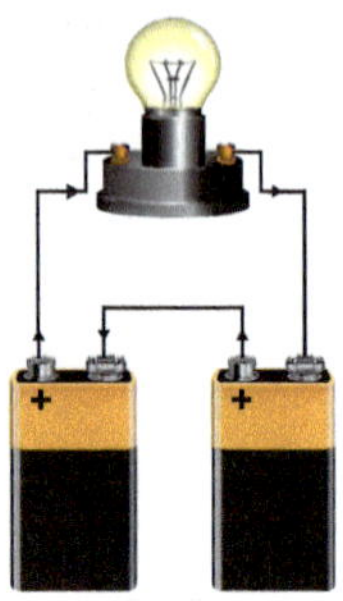

Battery
Resistance of the light bulb 3 Ω
Resistance of the motor 25 Ω

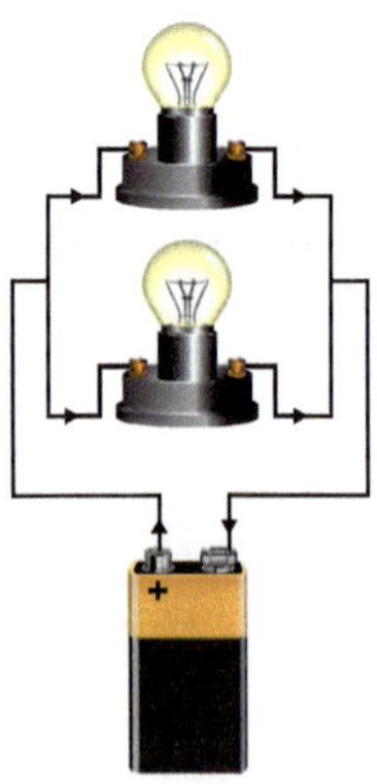

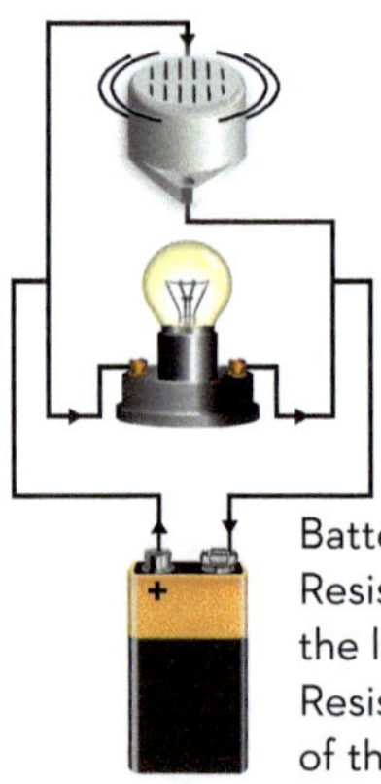

Battery 9 V
Resistance of the light bulb 3 Ω
Resistance of the bell 25 Ω

5. Calculate the current that circulates through the circuits in activity 4. In the circuits with elements in parallel, does the same current circulate through all the branches?

6. Represent the circuits in activity 4 with the corresponding symbols.

7. Why is it convenient to put a switch in all the circuits in activity 4?

8. With the circuit simulator that we used, build a circuit like the one in the figure. Vary the resistance of the light bulb, giving three values between 0 and 30 Ω. With the 'contactless ammeter', read the current that circulates. Interpret the speed of the particles of the current and the brightness emitted by the light bulb in each case. Compare your answer with your partner.

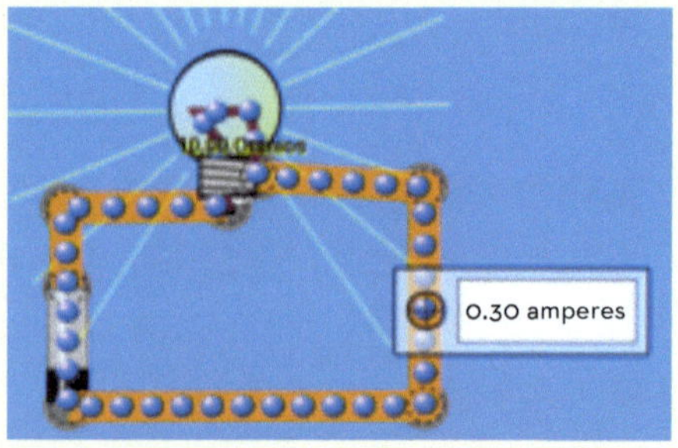

9. Could an electric circuit work without a generator? Why?

10. Look for information on how a fuse and a circuit breaker work. Then write five true/false statements about the fuse and the circuit breaker and give them to your partner.

11. With the help of your partner, research how electrical energy is transformed into electromagnetic energy in one of the old filament light bulbs. Does any other energy transformation occur?

12. In pairs, make a poster about a hydroelectric power plant near your city, including photos of the dam and an explanation of how it generates electrical energy.

13. Do you think there are more kilometres of roads or cables transporting electricity? Discuss with your partner.

14. Which way seems more logical to you to place the different receiving elements in a home, in series or in parallel? Explain your answer.

Consolidation activities

15. Work in pairs. What condition is necessary for there to be a flow of thermal energy between two bodies? What condition is analogous for there to be an electric current between two points of a circuit?

16. In an electric circuit, are the electrons that make up an electric current created by the generator?

17. In which of the two circuits shown will the light bulbs shine brighter?

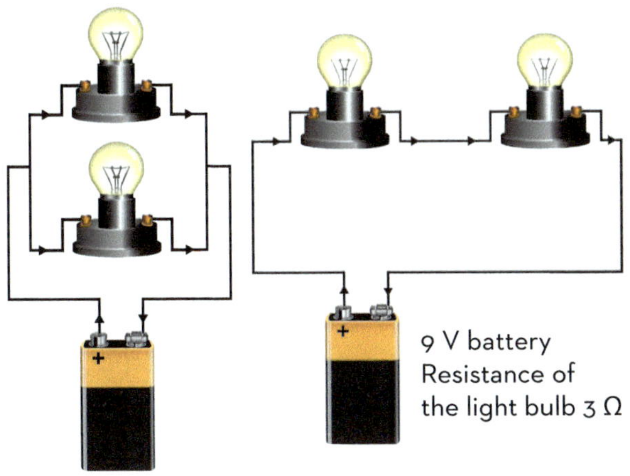

18. Use the circuit simulator and build a circuit with only a battery and cables that connect the two poles of the battery. Interpret what is shown in the simulator. Can a circuit function without receiving elements?

19. Why won't the LED in the circuit on the right light up? Note: An LED must not be connected to a 9 V battery because it would blow the LED.

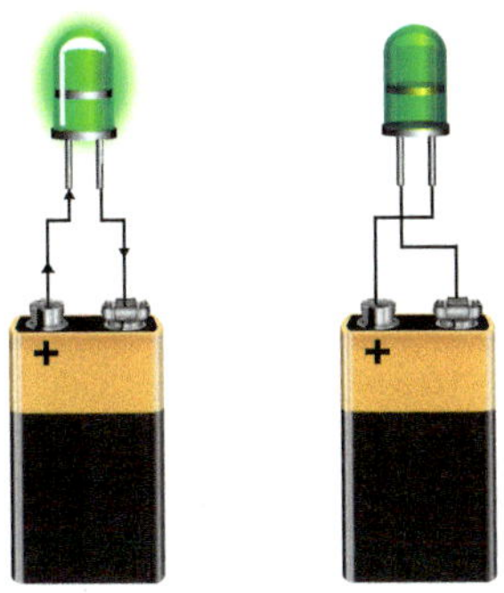

Advanced activities

20. Compare the functioning of a light bulb connected to a 220 V source with the same light bulb connected to a 120 V source. Which one is brighter?

21. With the help of your partner, look up information and prepare a slide show on how a speaker works. It converts electrical energy into sound.

22. Ammeters are devices for measuring current that must be connected in series. Use the circuit simulator to check what happens when you connect an ammeter in parallel. Is the resistance of an ammeter very high or very low?

23. In the following website: **www.unesa.net/unesa/html/sabereinvestigar/esquemas/esquemas.htm** you can see a diagram and an animated simulation of how the major electrical power plants work. Choose one of them, study the animation and explain it to your classmates. You can record a podcast or a video with your explanation

24. Think, pair up share: a friend tells you that he prefers a model of electrical device over another because it has the CE symbol and this label indicates the high quality of the product. Do you agree? What does the CE symbol mean? Share your answers with your partner and then with the whole class.

25. Explain how the following electronic circuit works. What use could it have?

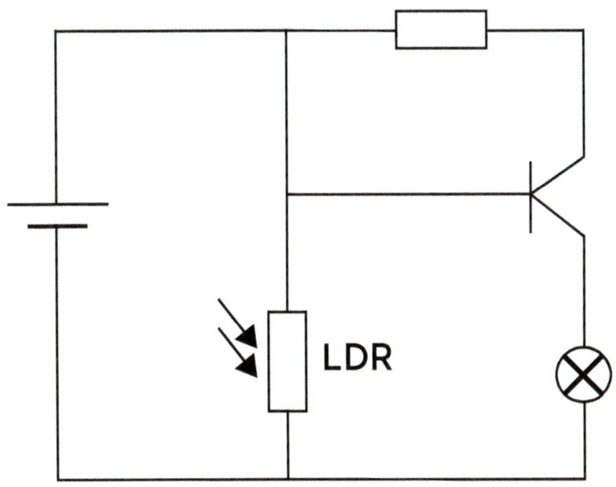

USE YOUR SKILLS

Blue electrical power plant

The figure shows a new type of electrical power plant located in a place where the fresh water of a river meets the salt water from the sea. The electrical power plant uses the different concentration of salt in the river water and the sea water to produce electricity.

In the blue electrical power plant, the fresh water from the river is pumped inside a tank. The salt water from the sea is pumped inside another tank. The two tanks are separated by a membrane that only water molecules can go through. The water molecules spontaneously go through the membrane, from the tank with a lower concentration of salt to the tank with a high concentration of salt. This increases the volume and pressure of the tank with the salt water, as shown in the detailed view (lower figure). This water then flows through a pipe and moves a turbine to generate electricity.

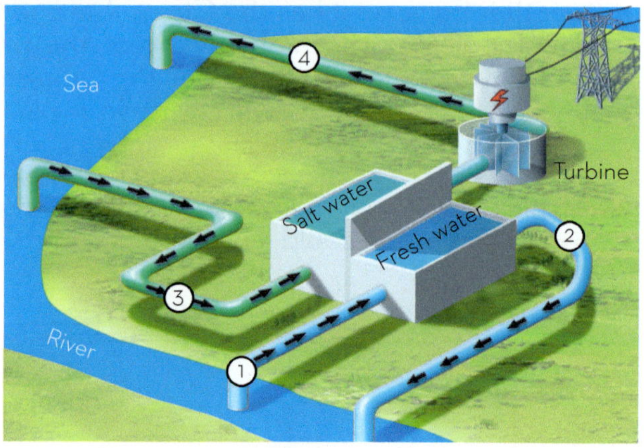

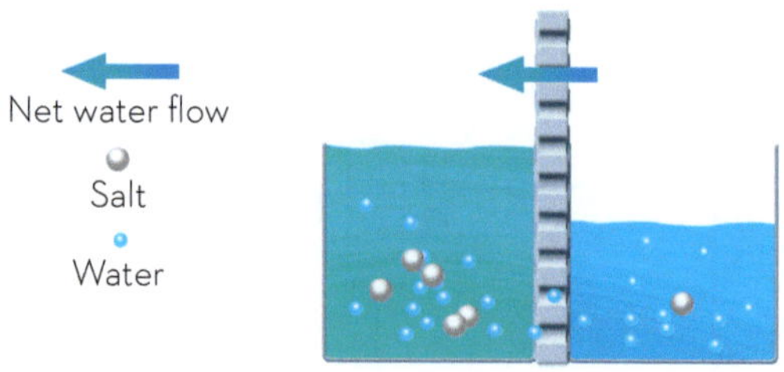

Take turns with your partner to answer the following questions:

Question 1

There are four numbered areas of the electrical power plant. Which areas of the process may have water molecules from the river?

Question 2

The river water has a low concentration of salt. When the molecules go through the membrane, does the concentration of salt in the fresh water tank increase or decrease? What about the concentration of salt in the salt water tank?

Question 3

What type of energy conversion occurs in the turbine and in the generator?

Question 4

Many electrical power plants use fossil fuels as an energy source, like coal or natural gas. Why is this electrical power plant considered more respectful with the environment?

Question 5

The energy produced in the power plant needs to be sent to towns located many kilometres away. What is the first step they must do with the voltage of the current generated?

Question 6

Imagine that the turbine produces 100 000 W (100 000 joules per second) of energy. The power plant pumps consume 10 000 W each. How many joules of electrical energy does this power plant produce every hour?

Organise an exhibit on the history of electronics

→I Competency task

Objective

The objective of this task is to organise a small exhibit on the **evolution** of **electronic devices**, and specifically on the process of **reducing** the **size** and **cost** of these devices and the transformation they have created in our lives.

Final product

The final product must have four scale models: three models of computers and one mobile phone. These models are significant in the history of computing. Each of these scale models must have an accompanying poster explaining the evolution from vacuum valves to transistors and finally to integrated circuits. You can see these components in the picture.

Step by step

1. Work in pairs. Look for information on the following four computer models: UNIVAC computer (1951), IBM 1401 computer (1959), Pentium computer from 1992 and a modern mobile phone.

   You should find information on the size of the devices, their price, their computing power or the type of tasks they are able to do, and the essential electronic component they have or had: vacuum valves, transistors or integrated circuits and the number of these components they have or had.

   You can also write down any other information about these devices that is interesting or impressive. For example: their mass, how many people were needed to use them, how information was entered and extracted, etc.

2. With this data, plan your exhibit. You should make a scale model of each one. A scale model means that if a computer is one hundred times bigger than another, the scale model must also be one hundred times bigger than the scale model of the smaller computer.

   The scale models can have the format you want. Think about the best way of demonstrating the miniaturisation of computers.

3. Design an explanatory poster for each model. The poster must be clear and fun to read. Do not use too much data or technical specifications. Only include important data that does not bore the reader. The poster must include the electronic component that was used in its design and a photograph of this component.

   The price of each device should also appear.

4. Make a final poster with the title 'Can You Imagine Daily Life with UNIVAC Computers'.

   This poster must reflect the enormous importance that miniaturisation and cost reduction of computers has had on our daily lives. On the web page 'Wikimedia Commons' you can find several pictures of these computer models (under Creative Commons Licenses) that you can use in your poster.

   Remember to quote your sources when you use information or pictures from the internet.

Self-assessment

Evaluate your work and your classmates' work using the guidelines that you can download from the OLC.

CHEMICAL FORMULATION

IMPORTANT

Currently, the **IUPAC** (International Union of Pure and Applied Chemistry) is the organisation in charge of maintaining the proper, methodical and precise language that constitutes the rules of nomenclature and formulation of the chemical compounds.

You have probably observed that when speaking or writing we use symbols (€, ∞, ©, $, &, kg, etc.) or initials (USA, EU, etc.) to represent longer words or concepts. We do this to simplify work and provide a clearer view of things. Chemists use the symbols of the elements to be able to write the compounds in a way that is easily recognised by other scientists, even if they are from different countries or cultures.

The symbols that identify the major elements are the following:

Hydrogen **(H)**	Manganese **(Mn)**	Silver **(Ag)**	Boron **(B)**	Aluminium **(Al)**
Lithium **(Li)**	Iron **(Fe)**	Cadmium **(Cd)**	Carbon **(C)**	Silicon **(Si)**
Sodium **(Na)**	Cobalt **(Co)**	Platinum **(Pt)**	Nitrogen **(N)**	Phosphorus **(P)**
Potassium **(K)**	Nickel **(Ni)**	Gold **(Au)**	Oxygen **(O)**	Sulphur **(S)**
Magnesium **(Mg)**	Copper **(Cu)**	Mercury **(Hg)**	Fluorine **(F)**	Bromine **(Br)**
Calcium **(Ca)**	Zinc **(Zn)**	Lead **(Pb)**	Chlorine **(Cl)**	Iodine **(I)**

The chemical language that identifies each substance has two expressions, the formula and the name.

The **chemical formula** identifies a substance with the symbols of the elements it has and the number of atoms of them.

The **name** of each substance identifies it and must have the proper form according to a set of common rules chosen by the scientific community.

WORKSHEET

Download this unit's worksheet and do activities 1 and 2.

For example, CO_2 is the formula and carbon dioxide is its name.

Chemical nomenclature is the set of rules that are used to name the different combinations of chemical elements.

Chemical elements may be found in their natural state in monatomic form (Fe, Na, Al, Cu, etc.), although they may also be found in diatomic form (H_2, N_2, O_2, F_2, Br_2, I_2, etc.) and even in polyatomic form, like phosphorus (P_4) and sulphur (S_8).

IMPORTANT

The symbols are written with one letter or if necessary, two, but the second is always lowercase. Note that Co is the symbol for cobalt, whereas CO is the compound carbon monoxide.

ACTIVITIES

1. **Some elements have symbols that are not related to their names. Look for the following ones: sodium, potassium, iron, copper, silver, gold, mercury, phosphorus and sulphur. Discuss with your partners where these symbols come from.**
2. **Work with a partner to explain why you think it is necessary for there to be rules for formulating and naming chemical compounds.**
3. **Look for the symbols of the noble gases and match them with their names.**

1 Compounds

Compounds are substances formed by a fixed and constant combination of different atoms.

Compounds are represented by a formula which indicates the number and the type of atoms that make up each molecule of the compound (if it is covalent) or the ratio between the ions (if it is an ionic crystal).

According to the number of different atoms in the molecular formula, the compounds are binary (NaCl, CO_2, etc.), ternary (NaOH, H_2SO_4, etc.) quaternary ($NaHCO_3$, etc.) and so on.

Now we will focus our study on the simplest compounds: binary compounds.

SCIENCE 2.0

You can distinguish elements from compounds on this website.

goo.gl/dZEqwX

2 Binary compounds

There are three types of binary compounds: oxides, hydrides and salts.

2.1. Oxides

These are combinations of oxygen with other elements. They end with the term **oxide**, indicating the atoms of oxygen and of the element with the prefixes **di** (2), **tri** (3), **tetra** (4), **penta** (5), etc.

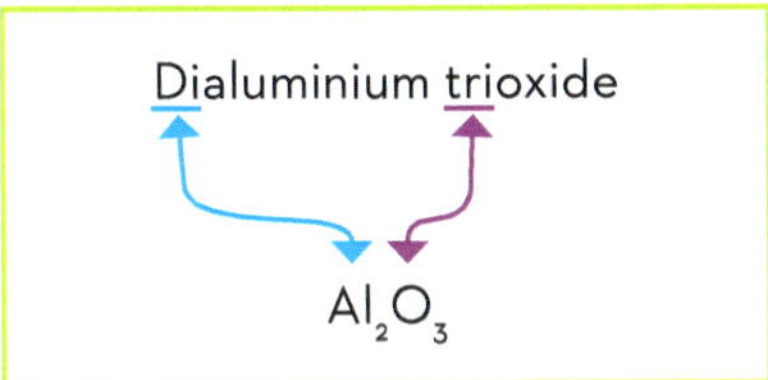

For example: CaO → Calcium oxide

Cu_2O → Dicopper oxide

Cl_2O_3 → Dichlorine trioxide

P_2O_5 → Diphosphorus pentoxide

DID YOU KNOW?

Dihydrogen oxide (H_2O) is water.

ACTIVITIES

4. Name the following substances:

Ni_2O_3	Fe_2O_3	FeO	PtO_2	MgO	I_2O_5

5. What is the formula for the following substances?

Sulphur trioxide	Nitrogen oxide	Silicon dioxide
Zinc oxide	Dibromine pentoxide	Dipotassium oxide

WORKSHEET

Download this unit's worksheet and do activity 3.

> **IMPORTANT**
>
> The prefixes used are:
>
> | | |
> |---|---|
> | 1 | no prefix or mono |
> | 2 | di- |
> | 3 | tri- |
> | 4 | tetra- |
> | 5 | penta- |
> | 6 | hexa- |
> | 7 | hepta- |

> **REMEMBER**
>
> Representation of the formulas of some chemical substances:
>
> 
> H_2O
>
> 
> CO_2
>
> 
> NH_3
>
> 
> N_2
>
> 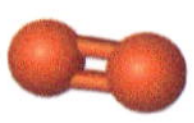
> O_2

2.2. Compounds with hydrogen

A. Hydrides

These are compounds of hydrogen with other elements. They end with the term **hydride**, indicating the atoms of hydrogen and of the metal with the prefixes **di** (2), **tri** (3), **tetra** (4), **penta** (5), etc.

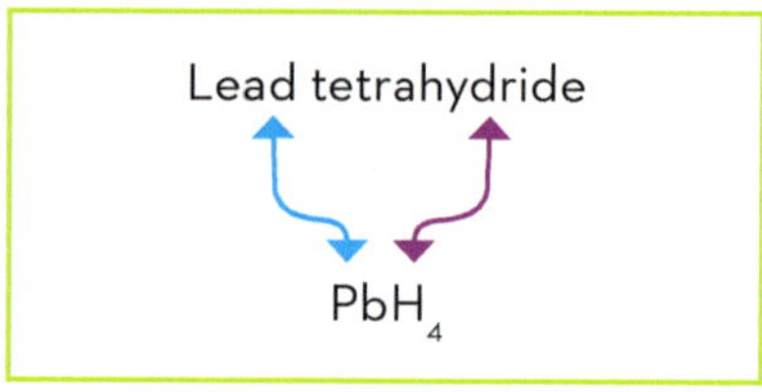

Some metal hydrides are:

Compound	Name	Compound	Name
CaH_2	Calcium dihydride	AlH_3	Aluminium trihydride

When there are bonds with non-metals (B, C, Si, N, P), they are named the same way. Some hydrides of this type are:

Compound	Name	Compound	Name
CH_4	Carbon tetrahydride	BH_3	Boron trihydride

When they are named as hydrides, the hydrogen atoms are written on the right side of the formula. For the rest, the hydrogen is placed on the left side of the formula.

Some non-metal hydrides have common names:

Ammonia	Silane	Methane	Phosphane	Borane
NH_3	SiH_4	CH_4	PH_3	BH_3

B. Hydrogen halides

When there are bonds with non-metals (F, Cl, Br, I, S), the atoms of hydrogen are indicated and then the name of the non-metal with the ending **-ide**. Some halides of this type are:

Compound	Name	Compound	Name
HCl	Hydrogen chloride	H_2S	Dihydrogen sulphide

The aqueous solutions (aq.) of the hydrides of F, Cl, Br, I, S are named hydracids and have the ending **-ic**.

Some solutions of this type are:

Compound	Name	Compound	Name
HF (aq.)	Hydrofluoric acid	H_2S (aq.)	Hydrosulphuric acid

2.3. Salts

These are compounds of metal elements with non-metal elements. They are named with the ending **-ide**, indicating the atoms of the metal and of the non-metal with the prefixes **di** (2), **tri** (3), **tetra** (4), **penta** (5), etc.

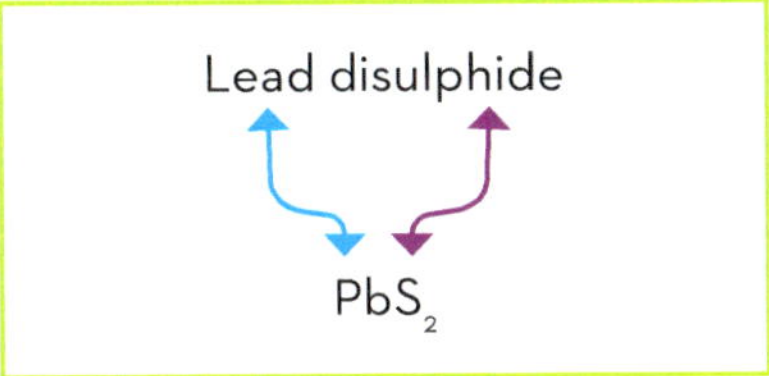

Some salts of this type are:

Compound	Name	Compound	Name
NaCl	Sodium chloride	SF_6	Sulphur hexafluoride

It is also possible to form salts by bonding non-metal elements with other non-metal elements:

Some salts of this type are:

Compound	Name	Compound	Name
CCl_4	Carbon tetrachloride	BI_3	Boron triiodide

SCIENCE 2.0

Here is a fun simulator for building molecules. How many can you make?

goo.gl/qS6E7u

ACTIVITIES

6. **Build the following chemical substances with jelly sweets. You will use the jelly sweets for the atoms of the elements and toothpicks for the bonds.**

   a) Copper dichloride. *b)* Ammonia. *c)* Magnesium dihydride.
   d) Platinum tetrahydride. *e)* Phosphine. *f)* Dialuminium trisulphide.
   g) Gold tribromide. *h)* Methane. *i)* Dimercury oxide.
   j) Dialuminium trioxide. *k)* Zinc difluoride.

   Also indicate which compound type we studied each of them belongs to.

7. **Name the following substances:**

NiH_3	Fe_2S_3	AuH_3	$PtCl_4$	KF	PbI_2

8. **What is the formula for the following substances?**

Dihydrogen sulphide	Ammonia	Calcium difluoride
Zinc chloride	Hydrobromic acid	Copper dihydride

WORKSHEET

Download this unit's worksheet and do activity 4.